Endemic Species from around the World - Teaching for Sustainability

Edited by Ana Cano Ortiz and Juan Peña-Martínez

Published in London, United Kingdom

Endemic Species from around the World - Teaching for Sustainability
http://dx.doi.org/10.5772/intechopen.1003342
Edited by Ana Cano Ortiz and Juan Peña-Martínez

Contributors
Aleksandra Talevska Leshoska, Ana Cano Ortiz, Camila Araújo, Danielle Viana, Diogo Nunes, Emmanuely Ferreira, Ilka Branco-Nunes, Juan Peña Martínez, Marina Talevska, Mary A. Colligan, Melissa L. Burns, Natalia Alves, Oumar Sadio, Patricia Yu, Paulo Oliveira, Sibele Mendonça, Suzanne L. Nelson, Trajche Talevski

First published in London, United Kingdom, 2025 by IntechOpen
IntechOpen is the global imprint of INTECHOPEN LIMITED, registered in England and Wales, registration number: 11086078, 167-169 Great Portland Street, London, W1W 5PF, United Kingdom

For EU product safety concerns: IN TECH d.o.o., Prolaz Marije Krucifikse Kozulić 3, 51000 Rijeka, Croatia, info@intechopen.com or visit our website at intechopen.com.

British Library Cataloguing-in-Publication Data
A catalogue record for this book is available from the British Library

Endemic Species from around the World - Teaching for Sustainability
Edited by Ana Cano Ortiz and Juan Peña-Martínez
p. cm.

This title is part of the Environmental Sciences Book Series, Volume 24
Topic: Ecosystems and Biodiversity
Series Editor: J. Kevin Summers
Topic Editors: Salustiano Mato, Josefina Garrido and Francisco Ramil

Print ISBN 978-1-83634-047-8
Online ISBN 978-1-83634-046-1
eBook (PDF) ISBN 978-1-83634-048-5
ISSN 2754-6713

If disposing of this product, please recycle the paper responsibly.

IntechOpen Book Series

Environmental Sciences

Volume 24

Aims and Scope of the Series

Scientists have long researched to understand the environment and man's place in it. The search for this knowledge grows in importance as rapid increases in population and economic development intensify humans' stresses on ecosystems. Fortunately, rapid increases in multiple scientific areas are advancing our understanding of environmental sciences. Breakthroughs in computing, molecular biology, ecology, and sustainability science are enhancing our ability to utilize environmental sciences to address real-world problems.

The four topics of this book series - Pollution; Environmental Resilience and Management; Ecosystems and Biodiversity; and Water Science - will address important areas of advancement in the environmental sciences. They will represent an excellent initial grouping of published works on these critical topics.

Meet the Series Editor

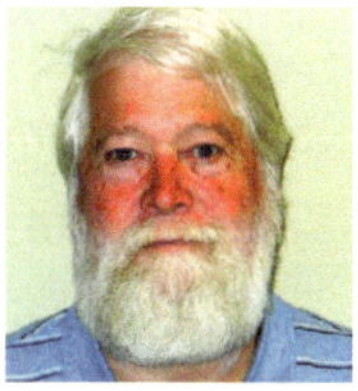

J. Kevin Summers is a Senior Research Ecologist at the Environmental Protection Agency's (EPA) Gulf Ecosystem Measurement and Modeling Division. He is currently working with colleagues in the Sustainable and Healthy Communities Program to develop an index of community resilience to natural hazards, an index of human well-being that can be linked to changes in the ecosystem, social and economic services, and a community sustainability tool for communities with populations under 40,000. He leads research efforts for indicator and indices development. Dr. Summers is a systems ecologist and began his career at the EPA in 1989 and has worked in various programs and capacities. This includes leading the National Coastal Assessment in collaboration with the Office of Water which culminated in the award-winning National Coastal Condition Report series (four volumes between 2001 and 2012), and which integrates water quality, sediment quality, habitat, and biological data to assess the ecosystem condition of the United States estuaries. He was acting National Program Director for Ecology for the EPA between 2004 and 2006. He has authored approximately 150 peer-reviewed journal articles, book chapters, and reports and has received many awards for technical accomplishments from the EPA and from outside of the agency. Dr. Summers holds a BA in Zoology and Psychology, an MA in Ecology, and Ph.D. in Systems Ecology/Biology.

Meet the Volume Editors

Dr. Ana Cano Ortiz holds a Ph. D. in Botany from the University of Jaén, Spain, with extensive experience in academia, private business, and secondary education. She has worked internationally in Spain, Italy, Portugal, and Central America, co-directing three doctoral theses and publishing over 100 papers in prestigious journals, books, and congresses. Her research focuses on botanical bioindicators, bioclimatology, agriculture, and science education applied to European, Central American, and Palestinian regions. As a professor in the Department of Didactics of Experimental, Social, and Mathematical Sciences, she is also a co-editor of a JCR journal and an active article reviewer. Dr. Ortiz combines scientific expertise with a commitment to innovative science education and environmental sustainability.

Dr. Juan Peña Martínez focuses on reshaping the training of future educators in the Didactics of Experimental Sciences, particularly in chemical education. His work integrates principles of human security, sustainable development, and peace education. He emphasizes core values such as the right to a dignified life, social justice, equality of opportunity, and the rejection of violence as a means of conflict resolution. His approach fosters generosity, dialogue, participation, and solidarity alongside the preservation of natural resources. Additionally, he seeks to cultivate knowledge, attitudes, and behaviors that promote environmental responsibility, addressing both ecological and human dimensions of sustainability.

Contents

Preface

In a world where natural boundaries are increasingly blurred by globalization, climate change and human activity, it is imperative to rediscover and value the endemic species that inhabit our planet. These species, unique to their region, are not only witnesses to the world's rich biodiversity but also guardians of ecosystems that sustain life as we know it. Yet many face an uncertain future, threatened by habitat loss, the introduction of invasive species and increasing environmental pressures.

Endemic Species from around the World – Teaching for Sustainability is born out of a profound need to generate awareness and action around these valuable biological treasures. This book is both a tribute to natural diversity and a practical tool for educators, students and environmental advocates who, through their actions and knowledge, can influence the preservation of our environment.

We have compiled information on endemic species representing the planet's diverse ecosystems in these pages. Each species has a crucial role in its habitat, and their preservation is key to the stability of local and global ecosystems.

However, this book seeks not only to inform but also to inspire. We advocate for transformative education that places sustainability at the center of curricula, fostering a deep respect for and commitment to nature in present and future generations. We believe that knowledge of endemic species, their ecological importance and the threats they face can catalyze a change in human mentality and actions.

We hope that *Endemic Species from around the World – Teaching for Sustainability* will be more than a reference text, we hope that it will become a tool for change, an inspiration for collective action, and a reminder that every effort, no matter how small, contributes to the greater goal of preserving the world's biodiversity.

Ana Cano Ortiz and Juan Peña Martínez
Department of Didactics of Experimental Sciences,
Social Sciences and Mathematics,
Complutense University of Madrid,
Madrid, Spain

Chapter 1

Introductory Chapter: The Role of Education in the Conservation of Endemic Flora Ecosystems

Ana Cano Ortiz and Juan Peña Martínez

1. Introduction

In order to conserve endemic and rare species, knowledge of the habitats, their mapping, their floristics and the relationship between these and man through educational services is a priority. European policy, through its Directive 92/43/EEC of 1992 on the conservation of fauna, flora and habitats of interest because of their endemic or rare character, establishes this. This European circumstance can be transferred to other places on the planet, especially to those areas with a high rate of plant endemism and a high floristic diversity. This is mainly due to the island effect, such as the islands of the Caribbean, Cuba, Española [1–3] and Madagascar, which act as hotspots on the planet [4], as well as in American areas where the Quaternary glaciations hardly caused extinctions, due to the special north-south disposition of the mountains, which favoured migratory routes, which is why in Mexico approximately 200 species of *Quercus* have been detected, bringing down the genus Pinus of holarctic character to subtropical environments [5]. Another factor that conditions the greater or lesser protection of endemic species is anthropic action; in this case, it is possible to intervene in a positive way to favour the conservation of the flora, which must be done by conserving the habitats; otherwise, the loss of genomes will continue, which in the opinion of Favarger and Contandriopoulos [6] represents genocide. Hence the need to stimulate and motivate the teaching of endemic flora in order to achieve its conservation, a plant diversity that inevitably requires the protection of the habitat in which it is located [7–9], habitats that are sometimes fragile due to their rarity or endemism. The greater or lesser abundance of rare and endemic species has its origin in geological, climatic, historical and edaphic aspects such as edaphisms [10]; of particular importance is the phenomenon of population isolation, which has not allowed genetic flow between individuals over time.

2. Results and discussion

2.1 Areas of distribution

Plants can be more or less stenotic, so there are plants restricted to small areas, while others have large areas of distribution, and their distribution is linked to the presence of certain ecological, historical and geographical factors. Among the various factors that condition the distribution of species, it is worth highlighting the ecological aspects, as can be seen in the endemic species of the south of the Iberian Peninsula belonging to the Asteraceae family [11].

Species present certain mechanisms to disperse their seeds, passing them from one place to another by anthropocoria, zoocoria, anemocoria, and if the seeds fall in biotopes suitable for germination, a new nucleus of propagation is generated, which is directly related to the stenotic nature of the plant. The boundaries of the areas have an ecological, geographical or biological component, being closely related to climate and soil. This interpretation is supported by the fact that most of the areas are situated within the floristic zones defined by temperatures (altitudinal floors) and by the degree of oceanicity [12–14]. For these reasons, the following areas of distribution exist.

1. Continuous areas. These are considered to be those which, despite the existing gaps, occupy the entire emerged lands or at least wide latitudinal spaces. In these cases, we speak of cosmopolitan occupation in the first case, and in the second case, depending on the area occupied, we speak of circumpolar, circumboreal, circumaustral and pantropical areas, which are generically included in circumterrestrial. Examples of cosmopolites are the reed *Phragmites* sp., cattail *Typha* sp. and duckweed *Lemna* sp., plants that are confined to marshy environments at almost all latitudes except in polar areas. Such dispersal seems to be attributable to migratory birds. A good example of cosmopolitan plants is the family Gramineae.

2. Disjunct areas. These are local areas without continuity, which is why they are grouped under the generic name of discontinuous areas. A case of discontinuity is when populations related at the species, genus or family level appear in areas sufficiently far apart, so that there can be no genetic flow between them. In this case, the phenomenon of speciation is occurring, since for any species there is the so-called disjunct threshold, which represents the minimum distance that cannot be exceeded by the species, so when we find areas above this threshold we are in the case of disjunct areas.

3. Vicariant areas. These are areas occupied by species that, although they have a common origin, due to a recent territorial separation (appearance of a geological accident), or because it is a very large area of occupation, there is no genetic exchange, and specific differentiations appear. These are cases of recent biological evolution, and generally only reach the subspecies level. This is the case of *Ulex parviflorus* and *Ules parviflorus* subsp. *eriocladus.*

4. Endemic areas. An endemic area is a very localised area which can have a very variable extension; the higher the rank of the taxon considered within the systematic scale, the greater the extent. Thus, species endemism is limited to a restricted area, such as a small mountain massif, *Viola cazorlensis*, *Lithodora nitida* of the Sierras Subbeticas and *Thymus mastichina* of the Iberian Peninsula; whereas genus, family and order endemism can extend to the whole of a continent, for example the American families Cactaceae and Bromeliaceae. Or this phenomenon is confined to islands where the island effect has led to high taxonomic differentiation; hence the high rate of endemic genera on islands such

as Madagascar and the Caribbean islands, for example Hispaniola has a rate of endemic species of more than 30%. Despite the different ways of classifying endemics, the most widely accepted classification is as follows: Paleoendemics, systematically isolated taxa, such as monotypic genera, as is the case of Tortuga Island, of calcareous nature, located to the north of Haiti, at a maximum altitude of 378 m. Despite its small size, the presence of the monotypic genus *Tortuella abietifolia* Urb. & Ekman and 15 exclusive endemic species justify Cano-Ortiz et al. to consider this small island as a district biogeographic unit [15]. These are ancient taxa, sometimes on the verge of extinction, making it a conservation endemism or relict endemism. Schizendemic, taxa that result from the slow and progressive differentiation (gradual speciation) of a primitive taxon in the different parts of its range by means of small mutations or recombinations. Schizendemic taxa have a common origin, since their formation is simultaneous and, being formed by gradual speciation, they have the same chromosome number. Patroendemics are taxa that have remained diploid in a given territory, while in neighbouring areas, they have given rise to corresponding polyploid taxa, the area of the latter being larger. Apoendemics are taxa that have originated in a given region by polyploidisation (sudden speciation) from a taxon of more or less large area and diploid or in any case of lower ploidy level.

2.2 Educational aspects

In the current situation, with a changing world and with great social and cultural challenges, it is necessary to move into a phase of action to protect endemic and rare species, and although most organisations in their resolutions talk about promoting environmental educational aspects, the reality is that these aspects always take second place. Consequently, it is a priority to raise awareness of the meaning of endemism in educational centres, as established by Noguera-Urbano [16]. Scientific literacy is a priority so that students and society can understand the value of species; otherwise, a large-scale destruction of species is foreseeable due to factors such as climate change, which is becoming more pressing every day, and the strong anthropic action that depletes plant resources. To promote botanical literacy, among other considerations, it is necessary to increase the cultural and educational level of the population with the learning of botanical-geobotanical concepts such as those set out by Cano-Ortiz et al. [17]; to this end, the few environmental contents taught in educational centres must be modified and increased, using teaching methodologies that have a greater impact on the acquisition of knowledge by students, either through enquiry [18–20], in any case the didactic proposal of practical classes outside the classroom by Álvarez and Antolin Rodriguez [21] is preferable, the latter method being essential as it arouses curiosity and develops the student's capacity for observation.

3. Conclusions

Although research on the knowledge of species has progressed in recent years, this is not the case at the sociocultural level of the population. It is therefore necessary to promote botanical studies in schools, educational centres and universities in order to raise public awareness, which is only possible with the involvement of public and private institutions.

Author details

Ana Cano Ortiz* and Juan Peña Martínez
Department of Didactics of Experimental, Social and Mathematical Sciences, UCM, Madrid, Spain

*Address all correspondence to: acano07@ucm.es

References

[1] Borhidi A. Phytogeography and Vegetation Ecology of Cuba. Budapest: Akadémiai Kiadó; 1991. 858 pp

[2] Borhidi A. Phytogeography and Vegetation Ecology of Cuba. 2ª ed. Budapest: Akadémiai Kiadó; 1996. 926 pp

[3] Liogier AH. La Flora de la Española. Vol. I-IX. In: Jardín Botánico Nacional Dr. Rafael Ma. Santo Domingo: Moscoso; 2000

[4] Cano-Ortiz A, Musarella CM, Piñar Fuentes JC, Pinto Gomes CJ, Cano E. Distribution patterns of endemic flora to define hotspots on Hispaniola. Systemtics and Biodiversity. 2016;**14**(3):261-271. DOI: 10.1080/14772000.2015.1135195

[5] Cano E, Velóz Ramirez A, Cano-Ortiz A. Phytosociological study of the *Pinus occidentalis* woods in the Dominican Republic. Plant Biosystems. 2011;**145**(2):286-297. DOI: 10.1080/11263504.2010.547685

[6] Favarger C, Contandriopoulos J. Essai sur l'endémisme. Vol. 9. In: Büchler, editor. Travaux de l'Institut de Botanique de l'Université de Neuchâtel. Institut de Botanique; 1961. pp. 1-25

[7] Cano-Ortiz A. Teaching about biodiversity from phytosociology: Evaluation and conservation. Plant Sociology. 2023;**60**(2):25-37. DOI: 10.3897/pls2023602/02. ISSN:2280-1855

[8] Gianguzzi L, Caldarella O, Campisi P, Sonia Ravera S, Scalenghe R, Venturella G. Plant diversity in old-growth woods: The case of the forest edges of the Favorita Park in Palermo (North-Western Sicily, Italy). Plant Sociology. 2024;**61**(1):1-29. DOI: 10.3897/pls2024611/01

[9] Raposo M, del Río S, Vazquez Pardo F, Cos JC, Cano-Ortiz A, Pinto Gomes C. New plant communities to define the southern boundary of the European Atlantic Province in mainland Portugal. Plant Sociology. 2023;**60**(2):39-55. DOI: 10.3897/pls2023602/03

[10] Cano E, Cano-Ortiz A, del Río S, Veloz A, Esteban Ruiz FJ. A phytosociological survey of some serpentine plant communities in the Dominican Republic. Plant Biosystems. 2014;**148**(2):200-212. DOI: 10.1080/11263504.2012.760498

[11] Cano E, Cano-Ortiz A, Musarella CM, Piñar Fuentes JC, Quinto-Canas R, Spampinato G, et al. Endemic and rare species of asteraceae from the southern Iberian Peninsula: Ecology, distribution and syntaxonomy. In: Tessier M, editor. Asteraceae: Characteristics, Distribution and Ecology. Serie: Plant Science Research and Practices. New York: Nova Publishers; 2020. pp. 147-175. Available from: https://novapublishers.com/shop/asteraceae-characteristics-distribution-and-ecology/

[12] Rivas-Martínez S, Fernández-González F, Izco J, Loidi J, Lousã M, Penas A. Vascular plant communities of Spain and Portugal. Addenda to the syntaxonomical checklist of 2001. Itinera Geobotanica. 2002;**15**(1):5-432

[13] Rivas-Martínez S. Mapa de series, geoseries y geopermaseries de vegetación de España. Parte I. Itinera Geobotanica. 2011;**18**(1):5-424

[14] Rivas-Martínez S. Mapa de series, geoseries y geopermaseries de vegetación de España. Parte II. Itinera Geobotanica. 2011;**18**(2):415-800

[15] Cano-Ortiz A, Musarella CM, Cano E. Biogeographical areas of hispaniola

(Dominican Republic, Republic of Haiti). In: Plant Ecology. London, UK, Croatia: Intech Publisher; 2017. pp. 165-189

[16] Noguera-Urbano EA. Endemism: Differentiation of the concept, methods and applications. Acta Zoológica Mexicana (n. s.). 2017;**33**(1):89-107

[17] Cano-Ortiz A, Piñar Fuentes JC, Ighbareyeh JMH, Quinto-Canas R, Musarella CM, Cano E. Geobotanical field activities for learning landscape interpretation concepts and methods for university students. Issue special "Didactic Experimental Science"-*Research*. Journal of Ecology and Environmental Sciences. 2022;**2**(2):11-22. DOI: 10.31586/rjees.2022.206

[18] Aguilera D, Martín-Páez T, Valdivia-Rodríguez V, Ruiz-Delgado A, Williams-Pinto L, Vílchez-González JM, et al. La enseñanza de las ciencias basada en indagación. Una revisión sistemática de la producción española. Revista de Educación. 2018;**381**:259-284. DOI: 104438/1988-592X-RE-2017-381-388

[19] Garritz A. Indagación: Las habilidades para desarrollar y promover el aprendizaje. Educación Química. 2010;**21**(2):106-110

[20] Taboada-Castro M, Lafuente Álvarez F, Turrión Nieves B, Martín Sanz R, Getino Doménech Girbau M, Marbà Tallada A. Gamificar la evolución: El ecosistema como contexto para la aplicación del pensamiento evolutivo. Revista Eureka sobre Enseñanza y Divulgación de las Ciencias. 2023;**20**(1). Available from: https://www.redalyc.org/articulo.oa?id=92072334016

[21] Álvarez M, Antolín Rodríguez JM. Propuesta didáctica para prácticas de Edafología y Climatología: Flipped Clasroom utilizando recursos audiovisuales. In: Redine, editor. 5th International Virtual Conference on Educational Research and Innovation. Proceeding CIVINEDU. Spain: Aldaya Press; 2021. pp. 486-489

Chapter 2

The Impact of the Maui Wildfires on Economic Sustainability, Public Awareness, and Environmental Stewardship in Hawai'i

Patricia Yu

Abstract

In 2023, the Maui wildfires caused extensive damage, burning over 2170 acres and destroying approximately 2207 structures. This paper examines the origins of the Maui wildfires, their economic impacts, and the shifts in public awareness towards wildfire risks and environmental sustainability. The Maui wildfires, driven by a combination of severe drought, strong winds, and downed power lines, significantly disrupted Maui's tourism-dependent economy, resulting in substantial property damage, business interruptions, and increased unemployment. Long-term consequences include depreciated real estate values and heightened food insecurity. This study highlights a growing public engagement in disaster preparedness and the integration of traditional Hawaiian ecological knowledge with contemporary wildfire management strategies. Restoration efforts emphasize sustainable land management, including invasive species control and community-based approaches to rebuilding. My findings underscore the necessity for proactive wildfire management, sustainable practices, and the integration of Hawaii's indigenous values to enhance resilience and ensure long-term recovery in Hawai'i.

Keywords: Maui wildfires, sustainable awareness, environmental stewardship, wildfires management, Hawai'i conservation

1. Introduction

In 2022, the United States experienced 68,988 wildfires, burning over 7.5 million acres, while in 2023, there were 56,580 wildfires, affecting more than 2.6 million acres [1]. In Hawai'i, approximately 0.5% of the land burns annually, predominantly due to human activities and the spread of non-native, fire-prone grasses in a dry, warm climate [2]. The 2023 Maui wildfires, which devastated western Maui, particularly Lahaina, resulted in an estimated $5.6 billion in damages. This tragic event showed the urgent need for effective wildfires management, sustainable economic practices, and improved environmental stewardship in Hawai'i. This study explores the Maui wildfires' origins, economic impact, and lessons for future sustainability. It also aims to analyze the economic impacts of the 2023 Maui wildfires and evaluate shifts in public awareness and attitudes towards wildfires risks and environmental sustainability. I will also discuss the implications for future environmental stewardship and sustainable development practices in Hawai'i.

1.1 Brief event summary

On August 8, 2023, the Maui wildfires ignited, driven by a combination of strong winds, dry conditions, and potentially downed power lines. This intense fire rapidly spread through the Maui island's leeward areas, consuming vast tracts of land and impacting both residential neighborhoods and natural habitats. Despite the swift response from firefighting units and community volunteers, the wildfires' progression outpaced containment efforts, leading to significant destruction and displacement.

1.2 Contributing factors

The severity of the Maui wildfires can be attributed to several factors including climate conditions, human activities, policy and management. The unusual dry spells and strong winds created favorable conditions for the wildfires' ignition and spread. And the expansion of urban and agricultural areas increased the availability of flammable vegetation.

Figure 1 is the meteorologic analysis of the 2023 Maui Wildfires. This figure shows that the Maui wildfires were influenced by an unusually strong high-pressure system. It also illustrates how strong easterly winds, enhanced by both the high-pressure system and local topography, can significantly impact fire spread. Such wind patterns likely played a role in the Maui Wildfires, where strong, dry winds could have exacerbated fire conditions. Although Hurricane Dora was too distant to directly

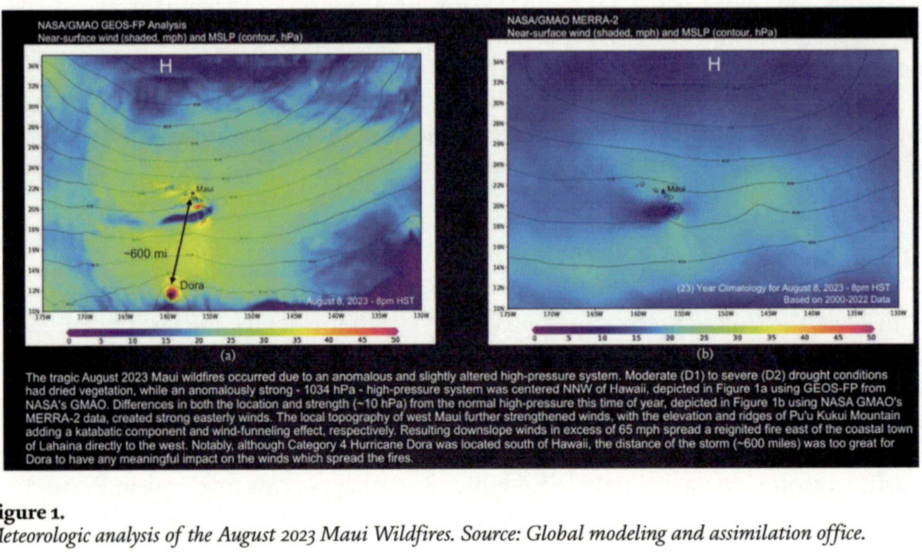

Figure 1.
Meteorologic analysis of the August 2023 Maui Wildfires. Source: Global modeling and assimilation office.

affect the winds significantly, its presence contributed to the overall weather pattern. This insight can help understand how even distant hurricanes might influence local weather conditions that contribute to wildfires. It reveals the role of an anomalous high-pressure system and local topographic effects in strengthening wind patterns. The comparison with climatological data indicates the deviation from normal conditions, reinforcing the significance of these anomalies in wildfire events.

2. Theoretical framework

The theoretical framework for this study is grounded in the interdisciplinary examination of environmental disasters and their socio-economic impacts, integrating theories from disaster management, environmental economics, and social resilience. The primary focus is on understanding how natural disasters, specifically wildfires, disrupt economic stability, public awareness, and environmental stewardship. Disaster Management Theory helps in understanding the phases of disaster management—preparedness, response, recovery, and mitigation—providing insights into how communities can prepare for and respond to wildfires, emphasizing the importance of risk reduction strategies and resilient infrastructures. Environmental Economics examines the economic impacts of environmental changes and disasters, exploring the cost of damage, recovery, and long-term economic consequences of wildfires on local economies, particularly in tourism-dependent regions like Maui. Social Resilience Theory looks at the ability of communities to withstand and recover from disasters, critical for understanding how public awareness and community engagement can enhance recovery efforts and promote sustainable practices post-disaster.

3. Methods

This paper utilizes a bibliographic review methodology, focusing on comprehensive analysis and synthesis of existing literature. A total of 30 documents were reviewed, covering various aspects of the economic, social, and environmental impacts of wildfires. **Table 1** outlines the research protocol for examining the impact of the Maui wildfires on economic sustainability, public awareness, and environmental stewardship in Hawai'i. This paper uses databases and sources such as MDPI, Elsevier, Science, Google Scholar, UHERO reports, Oxford University Press, Springer, and government reports. This research focuses on studies published between 2006 and 2024, using search terms like economic impacts, wildfires, environmental sustainability, public awareness, disaster management, and resilience. The included studies are empirical studies, systematic reviews, case studies, theoretical papers, government and non-government reports, news articles, and blogs. Inclusion criteria specify articles that assess the impacts of wildfires on economic sustainability, public awareness, and environmental stewardship, and are published in English. Exclusion criteria eliminate research not focused on wildfires or natural disasters. The geographical scope is global, with language limitations to English. Data extraction involves a standardized form detailing the author(s), year of publication, type of study, methodology, main findings, and conclusions. The analytical approach is narrative synthesis.

Items	Description
Research question	How do the Maui wildfires impact economic sustainability, public awareness, and environmental stewardship in Hawai'i?
Approach	Bibliographic Review
Databases and sources	MDPI [3], Elsevier [4], UHERO reports [5–9], Oxford University Press [10–12], PLOS ONE [13], Springer [14–16], Government Reports [1, 17–19], News Articles [20–23], University of Arizona Library [24], PNAS [25, 26], Science [27], UC Davis School of Education [28], East-West Center's Asia Pacific Bulletin [29], Hawai'i Wildfires Management Organization [2]
Search terms	Economic impacts, wildfires, environmental sustainability, public awareness, disaster management, resilience
Time frame	2006 to 2024
Types of studies included	Empirical studies, systematic reviews, case studies, theoretical papers, government and non-government reports, news articles, blogs
Inclusion criteria	Journal articles, government reports, news articles or blogs, studies assessed the impacts of wildfires on economic sustainability, public awareness, and environmental stewardship, published in English
Exclusion criteria	Research not focused on wildfires or natural disasters
Geographical scope	Global
Language limitations	English
Data extraction process	Standardized form including author(s), year of publication, type of study, methodology, main findings, and conclusions
Analytical approach	Narrative synthesis

Table 1.
Research protocol.

4. Economic impacts of the Maui wildfires

4.1 Immediate economic disruptions

Figure 2 illustrates the extent of the 2023 Maui wildfires damage. The burn area, highlighted on the map, shows the fire perimeter and the status of structures affected, with categories including destroyed, heavy damage, light damage, and undamaged. The estimated exposure includes 2719 structures exposed, 2170 acres burned, and 2207 structures damaged or destroyed, leading to a capital exposure of $5.52 billion, which is the estimated cost to rebuild. The estimated building exposure is categorized into 86% residential, 9% commercial, 2.4% educational, 1.1% unclassified, 1.0% industrial, 0.5% assembly, 0.4% agricultural, and 0.1% government buildings. The breakdown of potential needs includes support for 4500 sheltered population, 9000 meals ready to eat per day, 450 waste bins, 3560 gallons of water per day, and 51,700 square feet of shelter space. The inset image on **Figure 2** shows post-fire imagery dated August 10, 2023. Mengote [24] finds that the Maui wildfires caused significant interruptions in tourism, the primary economic driver in Lahaina, resulting in a sharp decline in visitor arrivals and extensive damage or complete destruction of local businesses. The wildfires also led to a surge in unemployment as numerous businesses

DOI: http://dx.doi.org/10.5772/intechopen.1006297

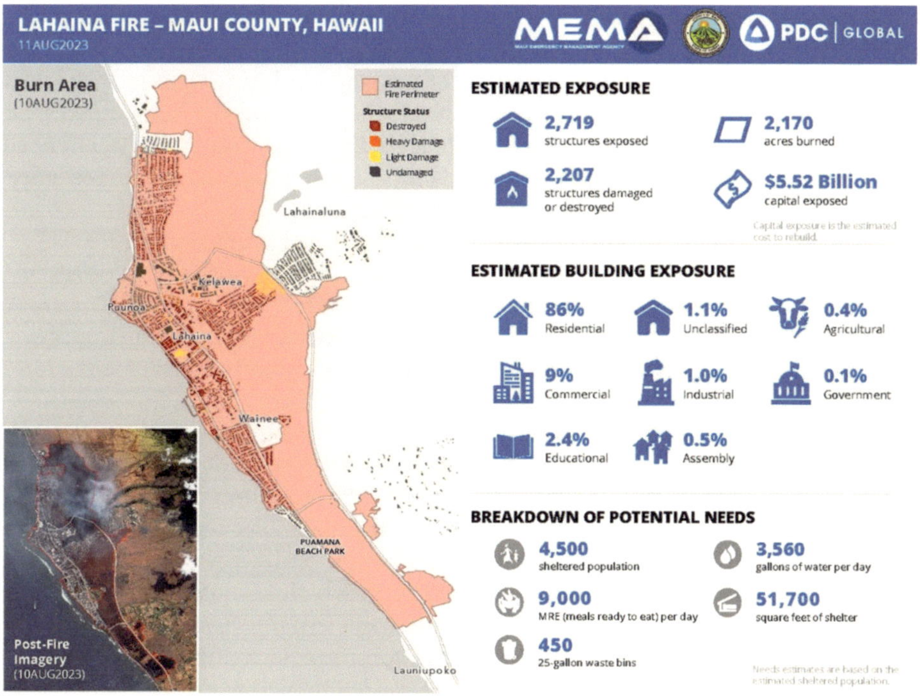

Figure 2.
The Lahaina fire damages in the Maui County, Hawai'i. Source: The Pacific disaster center and the federal emergency management agency.

and services were forced to close, exacerbating economic instability. The destruction of infrastructure and residential areas resulted in increased costs for emergency services, temporary housing, and initial rebuilding efforts [24].

4.2 Long-term economic consequences

Wildfires incur direct economic costs such as asset destruction and firefighting expenditures, but also generate substantial indirect costs by disrupting economic activities and depreciating asset values. The indirect and long-term costs of wildfires, which tend to accumulate over time, often surpass the immediate and direct costs. These indirect costs include decreased tourism revenue, reduced productivity in our economy, depreciated real estate values, increased healthcare expenditures, and the costs associated with post-fire ecosystem and residential rehabilitation [17]. Meier et al. [4] investigates the long-term economic consequences of wildfires in Southern Europe, revealing substantial Gross Domestic Product (GDP) losses due to both immediate damages and ongoing economic disruptions. Their revised analysis estimates annual GDP losses between €1.3 and €2.1 billion, correcting previous overestimated figures. This study [4] shows that wildfires result in prolonged economic impacts, affecting regional economies and employment, and highlights the necessity for accurate economic assessments to inform policy and enhance resource allocation for wildfires mitigation and recovery. Juarez et al. [6] provides an initial assessment of the long-term economic impacts of the Maui wildfires, focusing on the socio-economic and health consequences for affected residents. This study [6] reveals that the wildfires have significantly disrupted housing stability, with only 24% of participants remaining in their pre-wildfire homes. Economic repercussions include a 74% drop in household income, with 58% of participants losing their jobs and 24% still unemployed six months post-fire. The Maui wildfires have also heightened food insecurity, impacting 35% of households. Their findings underscore the urgent need for policies that address housing stability, healthcare access, and economic recovery to mitigate the long-term impacts of such disasters.

Tavor [14] examines the long-term economic impacts of wildfires on various sectors within the US capital markets, using a dataset of 161 wildfires from 2019 to 2022. This study [14] finds that wildfires significantly affect sectors such as insurance, real estate, and forestry due to direct damage and increased costs, leading to negative cumulative abnormal returns. Conversely, sectors like food and transportation may see positive returns due to increased demand and supply chain adjustments. Tavor's paper shows the varied responses across sectors and highlights the necessity of sector-specific risk assessments and mitigation strategies to address the complex economic repercussions of wildfires. Bonham et al. [5] provides a detailed analysis of the long-term economic impacts of the Maui wildfires, emphasizing its severe effects on various sectors of Hawai'i's economy. This report [5] outlines substantial disruptions in tourism, with visitor arrivals plunging by nearly three-quarters immediately following the tragic fire, resulting in significant losses in daily visitor spending. It projects a prolonged recovery for tourism, with complete normalization not expected within the forecast period. The Maui wildfires also led to high unemployment rates, particularly in Lahaina, with ripple effects causing widespread economic strain across Maui's local businesses. The rebuilding process is anticipated to be lengthy, affecting housing availability and construction sector dynamics, and posing challenges for the labor market. Crowley et al. [18] provides a comprehensive analysis of the economic impacts of wildfires in the United States, highlighting that the total costs of wildfires extend far beyond the commonly reported suppression expenditures. This report [18] estimates that wildfires impose annual costs ranging from tens to hundreds of billions of dollars, including direct damages, health costs, and various indirect economic impacts, such as reduced property values and long-term ecological degradation. It underscores the need for increased investment in wildfires management, including preparedness, response, and post-fire recovery, to mitigate these extensive economic burdens and enhance long-term resilience. Wibbenmeyer et al. [10] reviews the long-term economic impacts of wildfires on water quality, showing how these events alter watershed functions and increase water treatment costs. This study [10] identifies the significant but often overlooked economic burdens posed by degraded water quality due to wildfires, including increased costs for sediment removal, water treatment, and infrastructure repair. This paper emphasizes the need for comprehensive economic analyses to understand the full scope of wildfires' impacts on water resources, which are crucial for developing effective mitigation and adaptation strategies.

One of the primary impacts of the Maui wildfires on water quality is the increased sediment load in rivers and reservoirs. The intense heat from the fire can cause soil to become hydrophobic, reducing its ability to absorb water. This leads to increased runoff during rains, which carries ash, charred debris, and soil particles into water bodies. The loss of vegetation that typically stabilizes the soil further accelerates erosion, contributing to higher sediment levels. These sediments can clog waterways, reduce the capacity of reservoirs, and increase the need for dredging and water treatment. Wildfires also introduce a range of contaminants into water sources. The combustion of various materials releases pollutants such as heavy metals, polycyclic

Sector	Immediate impact	Long-term impact
Tourism	Significant decline in visitors	Prolonged recovery time
Real estate	Property damage	Depreciated values
Employment	Increased unemployment	Job losses

Table 2.
Economic impacts by key sectors.

aromatic hydrocarbons (PAHs), and other toxic substances. When these pollutants are washed into water bodies, they pose significant risks to both human health and aquatic ecosystems. In Maui, the presence of these contaminants has likely increased the cost and complexity of water treatment, as facilities must now remove a broader array of substances to ensure water safety. The reduced water quality can impact agricultural practices, as farmers may need to invest in additional filtration systems to use contaminated water for irrigation.

In summary, **Table 2** depicts the economic impacts of the 2023 Maui wildfires across key sectors: tourism, real estate, and job market. The immediate impact on the tourism sector is a significant decline in visitors, which has immediate repercussions on the local economy that heavily relies on tourism revenues. The long-term impact involves a prolonged recovery time, as the tourism industry will take an extended period to rebuild its reputation and regain the confidence of potential visitors. The real estate sector experiences immediate property damage due to wildfires. This damage translates into a long-term impact of depreciated property values, affecting the wealth and financial stability of property owners and investors in that region. The immediate impact on the job market is an increase in unemployment rates, as businesses affected by the wildfires are forced to close or downsize. The long-term impact includes job losses, which may persist as the local economy struggles to recover and rebuild, leading to sustained unemployment issues.

5. Public awareness and attitudinal shifts

5.1 Initial public reactions

The immediate aftermath of the Maui wildfires saw a surge in public awareness and concern about wildfires risks. The extensive media coverage raised awareness about the wildfires' impacts and the broader implications for environmental safety and efficient management. Yan et al. [20] reports on the initial public reactions to the devastating Maui wildfires of 2023, revealing widespread criticism and disappointment towards local officials for declining additional assistance before the fire escalated. This report [20] outlines significant failures in emergency preparedness and response, including the decision not to activate emergency sirens, which led to confusion and inadequate evacuations. Public sentiment has been characterized by frustration over perceived mismanagement and a lack of timely intervention, contributing to severe loss of life and property. Their findings shows the urgent need for improved disaster preparedness and more effective communication strategies to enhance public safety in future emergencies.

The positive reaction involves the local communities and government, who tried their best to provide aid and support, reflecting increased solidarity and proactive engagement in disaster response. The White House report details the extensive federal response to the Maui wildfire, showing significant community engagement efforts and support measures from the government [19]. Over 1000 federal personnel were deployed to Maui to assist with immediate response efforts, including search and rescue operations, provision of essential supplies, and coordination of sheltering for displaced residents. Financial assistance exceeding $12 million was provided to over 3300 households, supporting rental assistance and other needs. The establishment of FEMA's Disaster Recovery Center and the involvement of multiple federal agencies also facilitated the provision of housing support, healthcare, and economic recovery resources. Local organizations and residents also swiftly mobilized to provide essential support to those affected by the wildfires. The Hawaii Community Foundation established the Maui Strong Fund, aimed at offering financial assistance to the impacted individuals and families. The Hawaii Red Cross played a pivotal role in disaster relief, focusing on the immediate needs of shelter and supplies. Similarly, the Maui Food Bank's efforts ensured that the affected population had access to food, converting every dollar donated into four meals [22]. The Maui Humane Society demonstrated profound care for the non-human victims of the disaster, attending to pets that were lost or abandoned during the wildfires. They provided emergency medical care for animals suffering from burns and injuries sustained in the chaos. The Salvation Army Hawaii Division contributed by distributing boxed meals to those displaced, ensuring that the basic need for food was met during this crisis [22].

Community-based organizations (CBOs) were instrumental in addressing the vast, immediate needs, which included shelter, financial assistance, and debris removal. Their efforts were not limited to short-term relief but also extended to identifying long-term recovery strategies for the devastated communities in Maui. Volunteerism was another cornerstone of this island community's response. Local volunteers were essential at the Maui Mayor's distribution sites, where they helped prepare and distribute meals. The spirit of 'kokua'—helping others selflessly—was evident as residents of the Hawai'i state dedicated their time and resources to aid their neighbors [23]. The collective efforts of the Maui community exemplify the 'Aloha spirit'—a commitment to mutual respect and assistance. This spirit was the driving force behind the remarkable community-led recovery process, showcasing the power of unity in the face of adversity. This comprehensive response from both the public sectors and private sectors underscores the critical role of federal and local partnerships in addressing the diverse needs of the island community and fostering long-term resilience and recovery.

5.2 Educational initiatives

Following the wildfires, various educational programs were launched to enhance public understanding of wildfires risks and promote sustainable practices. Fu and Zhang [13] investigates how community-based educational interventions can enhance resilience against disasters by focusing on teacher resilience and well-being. Their study [13] highlights the crucial role of teachers in disseminating disaster knowledge and fostering a culture of preparedness within communities. They emphasize the integration of disaster education into school curricula, which improves both teacher and community resilience. Their review demonstrates that effective disaster education programs not only prepare communities for immediate responses to wildfires but also contribute to long-term sustainable practices and overall community resilience. This paper underscores the necessity of supporting teachers through training and well-being initiatives to maximize the impact of disaster education. Bonham et al. [7] explores the design and impact of disaster education programs implemented in response to the Maui wildfires. Their study [7] emphasizes the effectiveness of integrating traditional ecological knowledge with contemporary disaster preparedness strategies, promoting community participation, and employing culturally relevant educational methods. Key initiatives include community workshops, school-based programs, and interactive drills aimed at enhancing public understanding of wildfire risks and fostering sustainable practices. These programs have been successful in improving local resilience, encouraging proactive disaster management, and facilitating long-term sustainability. The authors argue that such educational efforts are vital for building a resilient community capable of effectively responding to future wildfires.

Khatibi et al. [15] investigates the effectiveness of public awareness, knowledge, and engagement in improving climate change adaptation policies. Their study [15] emphasizes the need for community involvement to enhance resilience against climate-related disasters. By reviewing literature on public participation, the authors identify that increased public knowledge and active engagement are crucial for the success of climate policies. They argue that integrating local knowledge with scientific insights, fostering community participation in decision-making, and using innovative educational strategies are essential for promoting sustainable practices and effectively managing disaster risks. They also highlight the role of public education in motivating communities to adopt sustainable practices and actively participate in disaster preparedness and response. For integrating local knowledge in Hawai'i, the Hawaiian indigenous knowledge, or ʻike Hawai'i, encompasses centuries of traditional ecological wisdom, practices, and cultural beliefs that can significantly enhance contemporary disaster management, particularly in fighting wildfires. This knowledge, deeply rooted in the understanding of local ecosystems, offers sustainable strategies for fire prevention, mitigation, and ecological restoration. Traditional Hawaiian land management, or ahupuaʻa system, divides the land into sections from mountain to sea, each managed to maintain balance and sustainability. This system can be adapted to modern fire management by promoting the cultivation of fire-resistant native vegetation that reduces the spread of fires. Native species like ʻōhiʻa lehua (Metrosideros polymorpha) and koa (Acacia koa), which have higher moisture content, can act as natural firebreaks. Integrating these plants into urban and rural planning can create buffers that slow fire progression and protect communities. Indigenous Hawaiians historically used controlled burns, or ʻōhiʻa lehua, to manage vegetation, reduce fuel loads, and promote the growth of specific plants beneficial for agriculture and cultural practices. These controlled burns, performed under precise conditions, can be reintroduced as a wildfire prevention strategy, mimicking natural fire regimes that maintain ecosystem health while reducing the likelihood of large, uncontrolled wildfires. The traditional Hawaiian knowledge of water management, including the construction of loʻi (irrigated terraces) and ʻauwai (irrigation channels), can be leveraged to enhance modern fire suppression efforts. These systems historically ensured water availability during droughts and can be adapted to support firefighting activities by providing strategic water sources across the landscape, thus enhancing the capacity to manage and suppress wildfires. The integration of indigenous knowledge into wildfire management also involves community engagement and education. Empowering local communities through educational programs that incorporate traditional knowledge and modern fire science can foster a collective approach to wildfire preparedness and response. This education can include training in traditional land stewardship practices, the benefits of controlled burns, and the importance of maintaining native vegetation.

5.3 Changes in public perception

Studies and surveys indicated significant shifts in public attitudes towards environmental issues. Steelman and McCaffrey [16] explores how public perceptions of wildfire management change following disasters when effective communication practices are employed. Through case studies of wildfires in California, Montana, and Wyoming, their paper [16] highlights that integrating risk communication before a wildfire and crisis communication during the event leads to greater public acceptance of adaptive fire management strategies. Their study underscores the importance of proactive education, transparent information sharing, and understanding the social context to transition from a suppression-only approach to more flexible fire management practices. Marlon et al. provides a comprehensive analysis of wildfire patterns in the western United States over the past 3000 years, highlighting the interplay between climate change, human activities, and wildfire risks [25]. Using sedimentary charcoal accumulation rates, the authors construct a historical record of fire activity and link it to climate data and human population dynamics. Their findings suggest that both climate variations (e.g., temperature and drought) and human interventions (e.g., fire suppression, land-use changes) have significantly influenced wildfire regimes. A notable observation is the 20th-century "fire deficit," attributed to effective fire suppression and ecological changes. This study underscores the critical role of climate in shaping fire regimes and points to the growing public awareness of the connections between increased wildfire risks, ongoing climate change, and anthropogenic impacts. This long-term perspective is essential for understanding current wildfire trends and predicting future fire activity under changing climatic conditions.

Bowman et al. [27] discusses the significant shifts in public perceptions following wildfires, particularly the increased awareness of the complex interactions between climate change, human activities, and disaster risks. This study [27] highlights how large, uncontrolled fires raise public consciousness about the role of human-induced climate change in exacerbating fire hazards. It also emphasizes that these disasters lead to a greater appreciation of the need for integrated fire management strategies that consider both ecological and socio-economic dimensions. Their research underscores the importance of fostering a holistic understanding of fire dynamics and promoting community engagement in developing sustainable practices to mitigate future wildfire risks. Cohn et al. [11] examines the evacuation behavior of rural communities during wildfires through three case studies in the western United States. Their research [11] reveals several key stages and dynamics of evacuation, including anticipation, warning, displacement, notification of the home's condition, and return and recovery. Individuals and communities tend to undergo significant behavioral changes due to the stress and disruption caused by wildfires. Some residents initially experience disbelief or denial, leading to delayed evacuations. During the displacement phase, evacuees prefer staying with friends or family over shelters, strive to maintain normal routines, and seek constant information about their homes. Their study also highlights the tensions between evacuees' needs for information and safety officials' responsibilities to maintain public safety and order.

6. Environmental stewardship and sustainable practices

6.1 Restoration and conservation efforts

Post-wildfires restoration efforts focused on rehabilitating affected ecosystems and promoting long-term sustainability. There are initiatives to replant native vegetation and restore natural habitats, which involves community volunteers and environmental organizations. Komenda and McAvoy discusses the ongoing efforts to restore culturally significant trees in Lahaina after the Maui wildfires, which are the deadliest in the U.S. in over a century, destroying vast areas of Lahaina, including ancient and culturally vital breadfruit (ulu) and kukui nut trees [21]. Arborists and volunteers are working to salvage these trees by digging for viable root matter and using it to propagate new growth. Restoration involves careful extraction of live tissue from burned trees to grow new shoots and propagate these important species. This effort is symbolically significant and is aimed at reconnecting the community with its cultural heritage. There are also efforts to reintegrate breadfruit into Lahaina, which include educating the community on its care and use, such as how to manage breadfruit trees in urban settings and utilize the fruit for food. This approach aims to make breadfruit a part of daily life and food security, rather than just a landscape feature. Reintroducing breadfruit, known for its versatility and nutritional value, could enhance food security for the local population, providing a sustainable food source amidst potential economic fluctuations. The restoration of these trees is seen to reclaim the cultural identity of Lahaina, connecting residents with their ancestral practices and landscapes. These efforts aim to revive historical food forests and integrate traditional agricultural practices into the modern community, which helps in ecological recovery and contributes to soil stability and the restoration of natural habitats, thus mitigating the environmental impact of the wildfires. These restoration efforts involve residents, arborists, and volunteers, revealing the community-driven nature of this project and fostering a collective effort to rebuild Lahaina.

6.2 Sustainable land management

To mitigate future wildfires risks and enhance sustainability, sustainable land management strategies are crucial. Nitta emphasizes the critical role of sustainable land management in reducing wildfire risks in Maui by integrating productive agricultural practices, diversifying land use, and controlling invasive species [30]. Maintaining actively managed agricultural lands helps lower fuel loads by preventing the proliferation of flammable invasive grasses, which are otherwise prevalent on idle fields and significantly contribute to wildfire intensity. Encouraging a variety of agricultural uses and crops enhances the resilience of both the land and Hawai'i's food systems, making the landscape more capable of withstanding and recovering from fires. Coordinated efforts to manage and eradicate invasive species are also essential to reducing the availability of tinder that exacerbates wildfires, thereby contributing to a comprehensive strategy for sustainable land management that mitigates fire risks. Synolakis and Karagiannis advocates for a comprehensive approach that integrates enhanced fire suppression with proactive mitigation measures, such as prescribed burns and fuel reduction, to manage fuel loads and mitigate catastrophic fire risks [26]. They emphasize the importance of aligning land management policies with fire risk reduction by managing biomass accumulation and adapting to climate-driven changes in forest ecosystems. Leveraging advanced technologies like satellite fire observations and real-time data analytics is crucial for improving wildfire predictions and emergency responses. Together, these strategies foster a proactive and integrated approach to wildfire management, balancing immediate fire suppression needs with long-term ecological resilience and technological advancements to enhance landscape sustainability.

Bond-Smith, et al. outlines a comprehensive approach to mitigating wildfire risks on unmanaged grasslands in Hawai'i by transitioning fallow lands into productive agricultural or ecological areas through improved fire codes and economic incentives [8]. They emphasize converting unmanaged grasslands, which harbor invasive species, into actively managed landscapes to reduce fire-prone vegetation and support local agriculture. The authors also advocate for integrating economic incentives, such as payments for ecosystem services, to encourage sustainable agricultural practices that align with fire risk reduction. By recognizing and compensating land managers for the ecosystem services they provide—such as reducing fire risks and enhancing biodiversity—this approach supports local food production and fosters resilient landscapes that balance ecological health with economic benefits.

6.3 Community-based approaches

Community involvement proved crucial in fostering environmental stewardship. Miller describes his project titled "Cultivating Youth and Community Resiliency: A Community Science Approach to Land Stewardship for Wildfire Mitigation in Maui, Hawai'i" [28]. This project integrates community science with land stewardship through a place-based, culturally responsive educational program. This initiative involves collaboration between Kanaka 'Ōiwi cultural practitioners, local ecologists, high school students from Kīhei Charter School, and educational faculty from UC Davis. Key activities include field investigations to manage invasive plant species and promote native vegetation, fostering a hands-on understanding of land stewardship and Hawaiian cultural practices among students. This program emphasizes the integration of Indigenous and Western knowledge systems to build community resilience and environmental stewardship. Expected outcomes include increased native plant abundance, reduced invasive grasses, and enhanced student engagement with land stewardship and cultural heritage. This approach highlights the effectiveness of culturally relevant education in addressing environmental challenges and promoting collective resilience.

Juarez et al. documents the wide-ranging impacts of the Maui wildfires on community health and social dynamics [6]. It emphasizes the role of community networks in disaster recovery, the exacerbation of pre-existing health disparities, and the critical need for continued support and resilience-building initiatives. Their study uses data from a diverse cohort of Maui residents to assess changes in health status, access to healthcare, and the effectiveness of community support systems post-wildfire. This important study is essential for understanding how community-based responses can effectively address the complex health and social challenges following natural disasters. It provides a model for integrating community feedback into disaster recovery planning and highlights the necessity of supporting local organizations to enhance community resilience.

7. Lessons learned and future directions

Several key lessons emerged from the wildfires, providing valuable insights for future wildfires management and sustainable development. One of the lessons is to utilize proactive fire management, which includes risk assessments, firebreaks, and community involvement. Building community resilience through education, resources, and engagement is crucial in mitigating the wildfires' impact and facilitating recovery.

Figure 3 depicts the status of areas covered by Community Wildfire Protection Plans (CWPP) across Hawai'i as of 2023. This map categorizes regions into three statuses: completed (marked in red), in process (marked in yellow), and national parks (marked in green). It shows that numerous areas across the islands have completed CWPPs, including North Shore, Western O'ahu, South Maui, Upcountry Maui, Western Maui, South Kona, North Kona, Hāmākua, and Ka'ū, among others, with completion years ranging from 2015 to 2022. Some areas, such as East Honolulu and Hāmākua, are in the process of developing their CWPPs. This figure shows that proactive wildfire management planning is in place in many parts of Hawai'i.

Figure 3 is relevant to the Maui wildfires' impact on sustainability because it illustrates the extent of wildfire protection planning in place prior to such events. Regions with completed CWPPs might have better strategies and resources for mitigating wildfire damage and enhancing resilience. Understanding the coverage and status of CWPPs can help assess the effectiveness of these plans in reducing wildfire risks and supporting sustainable recovery efforts. For Maui, which has several areas with completed CWPPs, this information is critical for evaluating the preparedness and response to the wildfires, and for planning future sustainability and resilience measures.

Sowby and Porter examines the intersection of water supply management and firefighting efforts during the catastrophic Maui wildfires, offering crucial insights

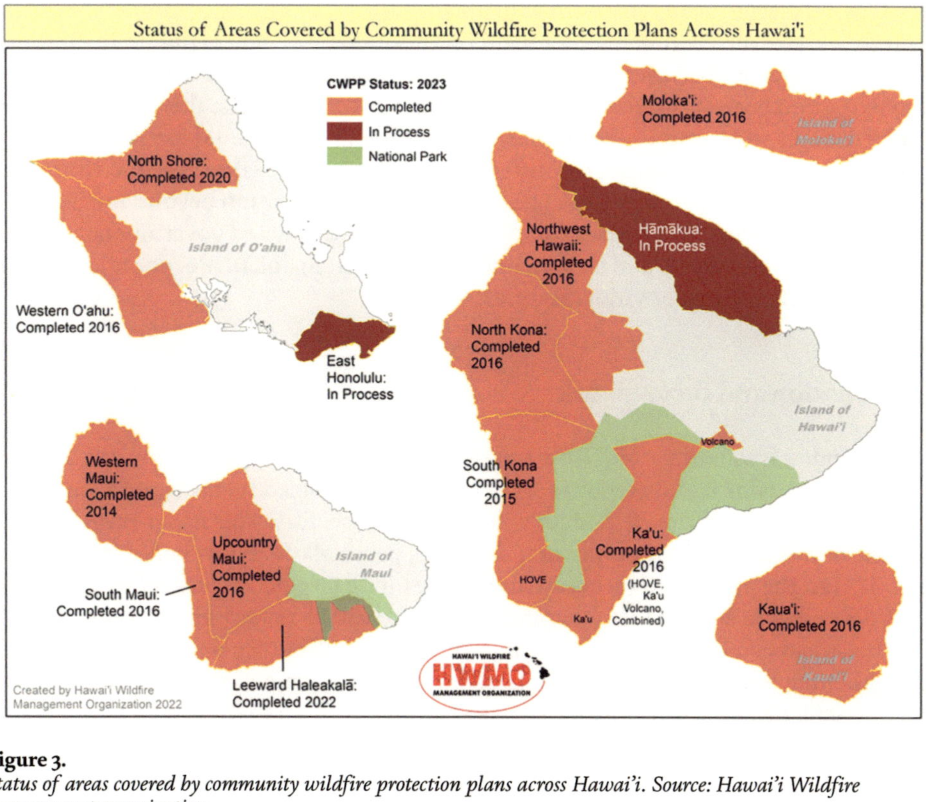

Figure 3.
Status of areas covered by community wildfire protection plans across Hawai'i. Source: Hawai'i Wildfire management organization.

into the limitations and necessary improvements for future disaster response [3]. Key lessons include the need for timely access to emergency water supplies, as delays significantly hindered firefighting efforts; the importance of backup power systems for water pumping, given the critical failures due to power outages; strategic prioritization of water use to prevent wasting resources on irrecoverable structures; and the implementation of protocols for shutting off water to destroyed buildings to preserve system pressure. Their findings underscore the urgent requirement for coordinated planning between water utilities and firefighting services, enhanced infrastructure resilience, and the integration of water management into wildfire preparedness strategies to better support community resilience in future events.

Arrasmith and Deptula analyzes the 2023 Maui wildfires, drawing critical lessons for enhancing disaster resilience [29]. The wildfires revealed significant vulnerabilities, including the inadequacy of aging infrastructure, such as utility poles, and the challenges posed by unmanaged invasive vegetation which exacerbated fire spread. Key lessons include the necessity of robust infrastructure upgrades to withstand extreme weather, improved land management practices to mitigate fire risks, and the importance of integrating modern fire-resistant technologies. Their report also highlights the need for better coordination between local and international partners to foster resilience, and the development of clear strategies to counter disinformation that may arise during crises. By addressing these issues, Hawaiʻi and other regions in the Indo-Pacific can better prepare for and mitigate the impacts of future disasters. Mak et al. explores the economic impacts and recovery strategies following the Maui wildfires [9]. Key lessons include the urgent need for resilient infrastructure and improved land management practices to mitigate future wildfire risks. The wildfires revealed vulnerabilities in Maui's dependence on tourism and the economic repercussions of disaster on a region highly reliant on this industry. They highlight the importance of integrating community participation in rebuilding efforts to ensure that reconstruction aligns with the needs and preferences of residents, thereby preventing displacement and promoting equitable recovery. Rebuilding should focus on creating more sustainable and resilient structures while balancing the island's economic recovery with long-term resilience strategies, including diversification beyond tourism. The recovery process emphasizes the critical role of coordinated government action, community involvement, and sustainable development to enhance economic stability and reduce future disaster vulnerabilities.

8. Discussion and future work

With more frequent natural disasters, there is a critical need for improved disaster preparedness and response strategies, as emphasized by the significant immediate impacts of the 2023 Maui wildfires across sectors. This paper reveals the high economic costs of environmental disasters, particularly in tourism-dependent regions, and the prolonged recovery times necessitate targeted economic support and investment in sustainable practices to mitigate future risks. Enhancing public awareness and community engagement is vital for effective recovery, suggesting that strengthening social resilience can help communities better cope with and recover from disasters. Future work can be focused on how to integrate Hawaiian indigenous values such as "aloha" (the spirit of "aloha" promotes harmony, kindness, and a deep respect for our environment and all living things, which encourages people to live with love and respect, fostering a sense of community and connection with others and with nature), "malama aina" (in Hawaiian culture, "malama aina" means "take care of the land", reflecting a deep connection to the land, recognizing that the health of the environment directly impacts the well-being of our community), "pilina" (in Hawaiian culture, "pilina" means relationship, stressing the value of fostering strong, positive relationships and maintaining harmony and balance in all aspects of life including our relationship with our environment), and "kuleana" (in Hawaiian culture, "kuleana" means responsibility, including the idea that individuals have a duty to care for the land, the community, and each other, reflecting a sense of stewardship and accountability) into disaster management strategies. Further research is needed to explore innovative risk reduction strategies and resilient infrastructures, as well as to assess the long-term socio-economic impacts of wildfires or other natural disasters on affected communities, ensuring a holistic approach to disaster preparedness and recovery.

9. Conclusion

The analysis of the 2023 Maui wildfires in this paper reveals profound insights into the impacts of such disasters and underscores the need for comprehensive strategies in wildfire management and sustainable development. The Maui wildfires, exacerbated by strong winds, dry conditions, and human activities, inflicted severe economic disruptions, long-term socioeconomic consequences, and highlighted significant gaps in emergency preparedness and response infrastructure. Economically, the immediate aftermath saw extensive property damage and business interruptions, particularly affecting Maui's tourism-dependent economy. The destruction of over 2000 structures led to significant losses in household income, increased unemployment, and heightened food insecurity. Long-term, the wildfires have diminished real estate values, disrupted tourism, and imposed substantial indirect costs on healthcare and ecosystem restoration. Studies emphasize that these impacts extend beyond immediate damage, affecting regional GDP and necessitating robust financial recovery measures.

Public awareness and attitudes towards wildfire risks and environmental sustainability have markedly shifted. The wildfires catalyzed increased community engagement and support for educational initiatives, aiming to enhance disaster preparedness and promote sustainable practices. Integrating traditional Hawaiian ecological knowledge with modern disaster management strategies has shown promise in fostering community resilience and effective land stewardship. Environmental stewardship efforts following the wildfires focused on restoration and conservation, including the replanting of native vegetation and the rehabilitation of affected ecosystems. Sustainable land management practices, such as controlling invasive species and promoting productive agricultural use of idle lands, are critical for mitigating future wildfire risks. In conclusion, the 2023 Maui wildfires have underscored the urgent need for improved wildfire management, sustainable economic practices, and enhanced environmental stewardship. Future strategies must include proactive risk assessments, community engagement, and comprehensive planning to build resilience against wildfires. The lessons learned from this disaster provide a roadmap for developing integrated approaches to mitigate wildfire impacts and foster sustainable development in Hawai'i and similar regions globally.

Author details

Patricia Yu
University of Hawai'i – West Oahu, Kapolei, Hawai'i, United States of America

*Address all correspondence to: pyu@hawaii.edu

References

[1] National Interagency Fire Center. Annual 2022 wildfires report. 2023. Available from: https://www.nifc.gov/fire-information/statistics/wildfiress

[2] Hawai'i Wildfires Management Organization. Hawai'i wildfires facts. Available from: https://www.hwmo.org/

[3] Sowby RB, Porter BW. Water supply and firefighting: Early lessons from the 2023 Maui fires. Water. 2024;**16**(4):600. DOI: 10.3390/w16040600

[4] Meier S, Elliott RJR, Strobl E. The regional economic impact of wildfires: Evidence from southern Europe. Journal of Environmental Economics and Management. 2023;**119**:102823. DOI: 10.1016/j.jeem.2023.102823

[5] Bonham CS, Gangnes B, Bond-Smith S, Fuleky P, Tyndall J, Inafuku R, et al. Wildfires Deliver Heavy Blow to Maui Economy. Honolulu, HI: University of Hawai'i Economic Research Organization (UHERO); 2023. Available from: https://uhero.Hawai'i.edu/wp-content/uploads/2023/09/23Q3_Forecast.pdf

[6] Juarez R, Maunakea A, Bonham C, Peres-David R, Phankitnirundorn K, Glatt N, et al. Initial Findings from the Maui Wildfires Exposure Study. Honolulu, HI: University of Hawai'i Economic Research Organization Public Health Report; 2024. Available from: https://uhero.Hawai'i.edu/wp-content/uploads/2024/02/InitialFindingsFromTheMauiWildfiresExposureStudy.pdf

[7] Bonham C, Bond-Smith D, Bond-Smith S, Bremer L, Burnett K, Coffman M, et al. Rebuilding resilience: Effective disaster education programs following the Maui wildfires. Journal of Disaster Research and Education. 2024;**15**(2):214-230. Available from: https://uhero.Hawai'i.edu/after-the-maui-wildfires-the-road-ahead/

[8] Bond-Smith S, Bremer L, Burnett K, Trauernicht C, Wada C. Reducing Fire Risk and Restoring Value to Fallow Agricultural Lands. Honolulu, HI: University of Hawai'i Economic Research Organization (UHERO); 2023. Available from: https://uhero.Hawai'i.edu/reducing-fire-risk-and-restoring-value-to-fallow-agricultural-lands/

[9] Mak J, Brewbaker P, Haas F. Economic Perspective of Maui's Devastating Wildfires. Honolulu, HI: University of Hawai'i Economic Research Organization (UHERO); 2023. Available from: https://uhero.hawaii.edu/blog/economic-perspective-of-mauis-devastating-wildfires/

[10] Wibbenmeyer M, Sloggy MR, Sánchez JJ. Economic analysis of wildfires impacts on water quality: A review. Journal of Forestry. 2023;**121**(4):374-382. DOI: 10.1093/jofore/fvad012

[11] Cohn PJ, Carroll MS, Kumagai Y. Evacuation behavior during wildfires: Results of three case studies. Western Journal of Applied Forestry. 2006;**21**(1):39-48. DOI: 10.1093/wjaf/21.1.39

[12] North MP, Collins BM, Stephens SL. Using fire to increase the scale, benefits, and future maintenance of fuels treatments. Journal of Forestry. 2012;**110**(7):492-401. DOI: 10.5849/jof.12-021

[13] Fu Q, Zhang X. Promoting community resilience through disaster education: Review of community-based

interventions with a focus on teacher resilience and well-being. PLoS One. 2024;**19**(1):e0296393. DOI: 10.1371/journal.pone.0296393

[14] Tavor T. Assessing the financial impacts of significant wildfires on US capital markets: Sectoral analysis. In: Empirical Economics. New York, NY: Springer; 2024. DOI: 10.1007/s00181-024-02574-3

[15] Khatibi FS, Dedekorkut-Howes A, Howes M, Torabi E. Can public awareness, knowledge, and engagement improve climate change adaptation policies? Discover Sustainability. 2021;**2**(18). DOI: 10.1007/s43621-021-00024-z

[16] Steelman TA, McCaffrey S. Best practices in risk and crisis communication: Implications for natural hazards management. Natural Hazards. 2013;**65**(3):683-705. DOI: 10.1007/s11069-012-0386-z

[17] United Nations Department of Economic and Social Affairs. Wildfires – A growing concern for sustainable development (Policy brief no. 111). 2021. Available from: https://www.un.org/development/desa/dpad/wp-content/uploads/sites/45/publication/PB_111.pdf

[18] Crowley C, Miller A, Richardson R, Malcom J. Increasing Damages from Wildfires Warrant Investment in Wildland Fire Management. Washington, DC: U.S. Department of the Interior; 2023. Available from: https://www.doi.gov/sites/doi.gov/files/ppa-report-wildland-fire-econ-review-2023-05-25.pdf

[19] The White House. Fact sheet: Biden-Harris administration's latest actions to support communities impacted by Maui wildfires. 2023. Available from: https://www.whitehouse.gov/briefing-room/statements-releases/2023/08/23/fact-sheet-biden-harris-administrations-latest-actions-to-support-communities-impacted-by-maui-wildfires-3/

[20] Yan H, Tucker E, Smart S, Von Quednow C. Maui Wildfires Report: Officials Declined Extra Help before a Deadly Inferno Engulfed Lahaina, Killing More than 100 People. Atlanta, GA: CNN; 2024. Available from: https://www.msn.com/en-us/news/us/maui-wildfiress-wind-swept-smoke-trapped-residents-and-impeded-evacuations-ag-s-first-report-says/ar-BB1lOs8N

[21] Komenda E, McAvoy A. The trees arrived with Polynesian voyagers. In: After Maui Wildfire, there's a Chance to Restore them. Tacoma, Washigntonz': Associated Press; 2023. Available from: https://apnews.com/article/lahaina-Hawai'i-maui-trees-wildfire-c35dd1e9fd63af23ca63a6eeb631c1d7

[22] Bjorn G. Maui Wildfires: A local's list of how to help. 2023. Available from: https://www.mauimaui.org/2023/08/11/maui-wildfires-how-to-help/

[23] Teague C. Help Maui fire victims: here's how you can donate. Civil beat. 2023. Available from: https://www.civilbeat.org/2023/09/help-maui-fire-victims-heres-how-you-can-donate/

[24] Mengote FM. Paradise on Fire: 2023 Maui-Lahaina Wildfire Case Study. Tucson, AZ: University of Arizona; 2024. Available from: https://repository.arizona.edu/bitstream/handle/10150/672304/MS-GIST_2024_Mengote.pdf?sequence=3

[25] Marlon JR, Bartlein PJ, Gavin DG, Long CJ, Anderson RS, Briles CE, et al. Long-term perspective on wildfires in the western USA. Proceedings of the National Academy of Sciences.

2012;**109**(9):E535-E543. DOI: 10.1073/pnas.1112839109

[26] Synolakis CE, Karagiannis GM. Wildfire risk management in the era of climate change. PNAS Nexus. 2024;**3**(5):151. DOI: 10.1093/pnasnexus/pgae151

[27] Bowman DMJS, Balch JK, Artaxo P, Bond WJ, Carlson JM, Cochrane MA, et al. Fire in the earth system. Science. 2009;**324**(5926):481-484. DOI: 10.1126/science.1163886

[28] Miller J. Cultivating Youth and Community Resiliency: A Community Science Approach to Land Stewardship for Wildfire Mitigation in Maui. Hawaiʻi: Citizen Science in Conservation Fellowship; 2023. Available from: https://education.ucdavis.edu/post/cultivating-youth-and-community-resiliency-community-science-approach-land-stewardship-wildfire

[29] Arrasmith P, Deptula C. Maui wildfires: Opportunities and lessons in resiliency for the Pacific region. In: Asia Pacific Bulletin. East-West Center; 2023. p. 656. Available from: https://www.eastwestcenter.org/sites/default/files/2023-11/APB%20656%20-%20Maui%20Wildfires.pdf

[30] Nitta N. In Fire-Stricken Maui, Sustainable Land Management is Key. Hudson, NY: Modern Farmer; 2023. Available from: https://modernfarmer.com/2023/09/fire-maui-land-management/

Chapter 3

A Framework for Implementing a Paradigm Shift toward a Proactive Approach for Conservation

Mary A. Colligan, Suzanne L. Nelson and Melissa L. Burns

Abstract

An examination of the conservation program in the Alaska Regional Office of the U.S. Fish and Wildlife Service revealed that changes in environmental conditions and corresponding changes in the timing and distribution of species were outpacing traditional conservation management methods. This led to a decision to shift the program more toward a proactive and collaborative manner, with less emphasis on utilizing a reactive approach. Efforts to shift the program included reducing staff workloads, increasing capacity, adding new skill sets, providing examples and a framework for proactive conservation, and building support from supervisors. Staff input and feedback was sought throughout the process and used to shift the culture of the program to foster strategic and collaborative conservation. An assessment of the proactive conservation program both provided encouragement and identified areas in need of additional attention. The current proactive conservation program has persisted through shifting agency priorities, declining budgets, and changes in internal leadership. The circumstances that necessitated a paradigm shift toward proactive conservation are not unique to Alaska; we urge others to consider implementation of proactive conservation or another paradigm that better aligns management approaches with the pace and scale of environmental change.

Keywords: proactive, conservation, paradigm shift, Alaska, framework, management

1. Introduction

Changes in environmental conditions with corresponding changes in the timing and distribution of species are outpacing traditional conservation management strategies. At a time when we can least afford delays in conservation, there is a dangerous mismatch between ecosystem decline and conservation methodologies currently in use. To address this misalignment, an effort was undertaken in the Alaska Regional Office of the U.S. Fish and Wildlife Service (Alaska USFWS office) to harness the expertise of biologists to chart and implement a new direction to improve conservation gains. The strategy that was used to address the current and foreseeable conservation challenges, termed proactive conservation, may provide insights for those considering similar initiatives or paradigm shifts within wildlife management and conservation organization offices.

Proactive conservation aims to use the best available science and collection of information that contributes to actionable conservation plans with achievable objectives [1]. This approach includes working in anticipation of threats to a species or habitat to mitigate or reduce the threats before the ecosystem becomes imperiled [2]. Proactive conservation encourages an ecosystem-based conservation approach in contrast to traditional management regimes that most often focus on individual species or featured locations and tend to utilize crisis-management [3]. Conservation strategies focused on supporting community-led efforts and promoting environmental values as a fundamental part of the management process are often included in proactive work [4, 5]. Proactive conservation planning can inform broad-scale land-use decisions by highlighting areas with high conservation value and assessing strategies that minimize risk to those areas [6]. Overall, proactive conservation work has been shown to be both more efficient and less expensive than traditional reactionary methods and policies, which often employ the delay-and-repair outcome that can be very costly over time [7–9].

Alaska offers an unequaled opportunity to practice proactive conservation across vast landscapes because it contains thriving fish and wildlife populations and habitats that are still relatively healthy and functionally intact. However, the state is not immune to changes that have occurred across the rest of the United States. Alaska is currently experiencing drastic environmental change, and initial signs of habitats and species under pressure are becoming readily apparent [10, 11]. For example, Alaska is undergoing climatic warming at more than twice the global rate, resulting in shifting biomes, changing ecological functions, and a redistribution of species [12–14].

There is added complexity in planning around uncertainties of future climate change [15]. Ecosystems are seen as complex, adaptive systems with multiple possible trajectories [16]. Historic conditions were often used as a management benchmark; however, rapid climate change has caused ecologists and natural resource managers to reconsider whether desired ecological conditions [10] goals should be historically based, based on the current set of conditions, or future-based addressing how to adaptively manage ecosystems in transition [14]. Recovery or restoration to a previous state may not be achievable or even desirable considering the way ecosystems and environments are shifting from historical baselines that are generally observable, knowable, and agreed on to nonstationary conditions that are new, unknown, and contested [17, 18]. Instead, a more biologically relevant goal may be to increase the resiliency of species and their habitats in anticipation of future climate challenges. Proactive conservation work has the potential to develop conservation strategies informed by and tailored to work within future climate change trajectories.

We provide a conceptual framework for a paradigm shift toward a greater emphasis on proactive conservation within the Alaska USFWS office. Conservation challenges faced by this office are not unique, nor are the obstacles encountered when attempting to implement a paradigm shift. We detail how the concept of proactive conservation was conceived, developed, and implemented in the Alaska USFWS office and strive to explain the process in a general way so that the lessons learned may assist other conservation entities seeking to implement similar paradigm shifts.

2. Methodology

The process by which the paradigm shift was implemented in the Alaska USFWS Office was both iterative and adaptive and was comprised of multiple steps (see **Figure 1**). In 2016, efforts began to engage staff in a program-wide strategic planning process to solicit input and opinions on what projects occupied their time, which projects they preferred to focus their efforts, and what actions and projects would have the greatest conservation benefit. The strategic plan that emerged from this review was seen as an important keystone document that provided a clear statement of: (1) the program's future direction; (2) commonly held values by staff and mangers; (3) a commitment to a desired work environment; (4) consistent criteria to establish work priorities; and (5) a commitment to implement streamlining procedures to provide staff the time and resources to dedicate to proactive planning and landscape-level efforts.

A key component of the strategic plan was a commitment to create and foster a work environment that emphasized risk analysis, encouraged innovation, and rewarded courage. Inclusion of this commitment was, and continues to be, controversial. In informal conversations, many staff expressed skepticism that they could be innovative or courageous in their work and seemed uncomfortable when asked about their work relative to this stated commitment. Management recognized that including the words risk analysis, innovation, and courage were not the norm for a government document; their inclusion was intended to provoke a reaction. The thought behind this novelty was that doing things the way they had always been done was not maximizing conservation benefits, so the desire was to push the system toward what is referred to as the productive zone of disequilibrium. The productive zone of disequilibrium is defined as the optimal range of distress within which the urgency in the system motivates people to engage in adaptive work. In other words, this is the place where people are creative and develop novel solutions [19].

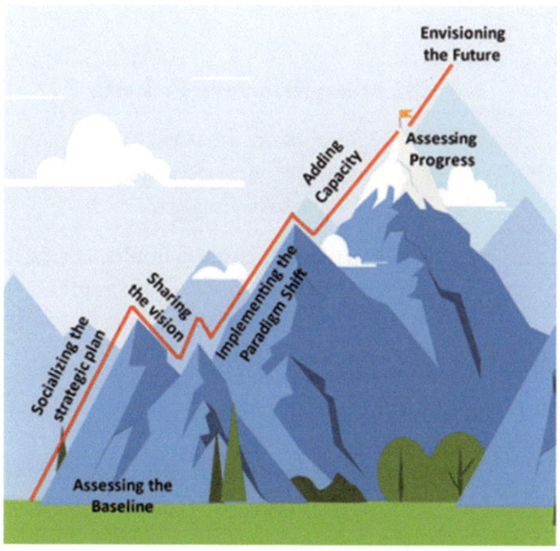

Figure 1.
Diagram of the phases implemented in the Alaska U.S. Fish and Wildlife Service office to shift from the traditional reactionary conservation approaches to a proactive conservation framework. Phases include assessing the baseline, socializing the strategic plan, sharing the vision, implementing the paradigm shift, adding capacity, assessing progress and envisioning the future.

During the baseline assessment, most staff identified an imbalance between the work where they spent most of their time, which often had very limited conservation value, and the work that they wanted to do, which often had greater conservation value. However, at the same time, they expressed hesitation in making an abrupt change in how they approached their work, which indicated to management that staff needed both a transition process and time for the shift toward work with greater conservation value to be successful. As a result of this analysis, the strategic plan prioritized staff spending more time and resources on proactive planning and landscape-level efforts given their potential to provide greater conservation impact. These priorities are grounded in the belief that conservation is best achieved in a proactive and collaborative manner, such as by crossing program and office boundaries, rather than by solely utilizing reactive and project-by-project methodologies.

An opportunity arose in the summer of 2018 to add additional capacity to assist with changing the culture of the program to one of greater ownership and engagement by staff. It was recognized that the program faced the adaptive challenge of changing the culture of the program to foster strategic and collaborative conservation [20]. Communication with staff was the focus of this phase and involved interviewing 28 individuals, observing the work of the team, and seeking feedback and input. The themes explored in these interviews are presented in **Table 1**. The purpose of the interviews was to gain insights from staff on their perceptions of the existing conservation program, prompt their thinking about how the program could be different under a new paradigm, hear what they saw as barriers to change, and to learn how the process used to facilitate the shift was working for them. A report developed following the staff interviews identified the need for clearer guidance from leadership so that staff could meet expectations for their work. The report described a significant amount of uncertainty among staff but portrayed a committed and engaged workforce dedicated to conservation.

In response to the request from staff to reduce uncertainty around the proposed paradigm shift toward proactive conservation, a vision document was drafted and included the expectation that staff dedicate 20% of their annual work time to projects

Key steps	Example questions for each step
Assessing the baseline	1. In your opinion, what are we doing now that will have lasting impacts? 2. What kind of work do you want to be doing that you think would have a larger conservation impact?
Envisioning the future	3. What are your ideas on implementing the proactive conservation approach in our office and program? 4. What does your program look like when proactive conservation is successfully implemented?
Identifying barriers and linkages	5. What are the barriers that prevent you from working proactively in your everyday work? 6. What linkages across programs/offices do you envision could help with a more proactive approach to conservation?
Assessing the process to support the paradigm shift	7. Did you feel invited to participate in the process to the extent you were interested? If not, why not? 8. What is working for you in this process? What is not working?

Table 1.
Key steps to launching a paradigm shift within an organization or group and example questions for each of the key steps as part of the paradigm shift toward proactive conservation in the Alaska USFWS office.

DOI: http://dx.doi.org/10.5772/intechopen.1006502

focusing on proactive conservation. This, coupled with the development of best management practices to streamline current workload, was intended to create space in employee's daily work activities to accommodate a shift toward proactive conservation. The inclusion of proactive conservation to staff's work time was intended to make the commitment to this paradigm shift tangible and real for staff and supervisors. The value 20% was chosen to provide sufficient time for the work to be meaningful so that the employee would feel a difference in their work plan and so that conservation results could be realized. It was intentionally left vague as to how the 20% could be achieved to allow flexibility—it could be 1 day a week, 1 week a month, or the time could be concentrated during a field season as long as it amounted to 20% of the total annual work hours of the employee. The overall intention was to provide both expectation and accountability, but also to empower staff to take the initiative to chart their own course toward greater engagement in proactive work.

To gain additional capacity, provide fresh perspective and create a confidential and judgment-free opportunity for staff engagement, leadership sought outside support to explore proactive conservation concepts and application in a separate and second round of staff interviews. This approach had the added advantage of reinforcing that this was not a top-down mandate, but instead grew directly out of the feedback provided by staff. The opportunity for open dialog with someone with applied expertise and an outside perspective, and the commitment to include staff input in a final report and presentation to leadership distinguished the proactive program from past efforts.

Staff and managers from multiple subprograms and locations within the Alaska USFWS office were provided the opportunity to discuss the concept of proactive conservation in interviews lasting approximately 1 hour (see **Table 1**). The questions were not sent ahead to the staff or managers so that first impressions and thoughts could be captured during the interviews, rather than prepared responses. Questions were purposefully open-ended to encourage more conversation.

A total of 31 interviews were conducted, representing the breadth of programs within the office. This was done with the intention of providing a diversity of views from those already involved in proactive work to those voicing skepticism for the proactive conservation concept. In addition to the many themes found within the responses, interviews were also instrumental in generating new ideas for proactive work. Program staff and managers identified the biggest barriers to engaging in proactive work as a lack of time and capacity, lack of a framework for how to go about the work, unknown support from supervisors, and a hesitancy in collaborating with some groups within the program. In response, pilot projects to demonstrate both what proactive work could look like and how it could be undertaken were developed. The intention was to demonstrate how to identify opportunities for proactive work and the appropriate steps to build the necessary partnerships. The point of this pilot project was to illustrate *how* to go about the process of developing proactive conservation projects and develop a framework and examples for others to follow.

For there to be a real shift toward proactive conservation, the concept needed to be incorporated systemically. One of the most effective ways to implement and sustain change is to integrate it into budget and workforce planning. While the goal remained to have all staff spend time on proactive conservation, it was recognized that the transition would take time, and there was a risk that the initiative would lose its momentum without someone actively and continuously endorsing the concept. A commitment was made to create and fill two full-time, permanent positions to maintain momentum and further catalyze and implement the proactive conservation initiative. Major priorities of a newly created Regional Proactive Conservation Coordinator included building relationships with other conservation agencies, identifying and pursuing proactive conservation opportunities with internal and external partners, and providing guidance and assisting staff on how to engage in proactive conservation work. To build additional capacity, a Proactive Conservation biologist was hired and placed in a program where there was the greatest need to build bridges to other programs and create linkages between offices to realize conservation opportunities. Additional capacity for the proactive concept came from a biologist contractor whose primary duties included identification of opportunities, development of outreach information and educational resources, partnership development, and meeting facilitation and support.

In addition to hiring these positions, hiring panels for all new positions discussed the proactive conservation paradigm when interviewing new applicants. Initiating this conversation at the hiring stage ensured that any individual coming into the program arrives with the expectation and understanding that proactive conservation is an important part of the program. Some candidates provided feedback that this new practice of discussing proactive conservation made the position and program more attractive.

Significant portions of the program budget are allocated to foster the development of proactive conservation projects corresponding to programmatic priorities. Directing funding to the implementation of proactive conservation work demonstrates a commitment to achieving programmatic objectives and supporting opportunities within the program, including supporting projects that a single program may not be able to accomplish alone. This financial commitment allowed for the development of a request for proposals and created a financial incentive and leveraging opportunities to foster the development of projects promoting proactive conservation.

3. Results

The proactive conservation program needed to be adaptively managed if it was going to successfully develop and transition to realize its full potential. Therefore, a commitment was made to regularly assess the program including conducting an evaluation 5 years after inception and 12 months after implementation, and to modify the program as needed to determine if the methodology employed had resulted in the desired paradigm shift. In February of 2022, a Proactive Conservation Program Assessment (hereafter Assessment) was distributed to all staff within the program. The Assessment response rate was 39%, with 50 of 122 individuals employed within the program responding voluntarily. The Assessment was intended to evaluate the effectiveness of the actions taken to address the barriers staff previously identified, and to identify additional steps necessary to achieve the proactive conservation paradigm shift.

All the respondents indicated they had interest in proactive work, and 60% of respondents indicated that they were already doing proactive work. Many respondents (56%) indicated that proactive conservation was explicitly included in their current work plan. These responses were very encouraging, and a good indicator of success for the first year of the program and demonstrate that proactive conservation is now generally embraced by staff.

The Assessment also asked respondents to use one word or a phrase to describe their feelings toward proactive conservation. These responses were used to generate

Figure 2.
Word cloud developed from 50 responses from polled staff during the proactive conservation program assessment in response to the prompt, "In a few words, describe your thoughts about proactive conservation." This figure was designed using WordItOut.com.

a word cloud, where larger words represent more frequently referenced ideas (**Figure 2**). The snapshot provided by the Assessment indicates that the level of awareness of the proactive conservation program has increased, as has the number of staff members engaged in proactive work. The Assessment revealed differences in attitudes toward proactive conservation from the staff to mid-level managers to program leaders. While the proactive conservation program appears to be broadly accepted and endorsed at the staff-level, there is an identified resistance to program implementation by mid-level managers.

The Assessment asked respondents to rank what tools or additional support was needed for staff to be able to work more proactively. Staff prioritized the need for direction from leadership and expanded messaging, a framework to guide program implementation, and multidisciplinary teams to brainstorm and scope projects. Approximately an equal number of respondents ranked relief from existing workload as their first choice as those who ranked it as their last choice. Interestingly, awards to incentivize and recognize proactive work were not identified as a need, with 50% of responding staff ranking it as the lowest priority.

Two significant transitions in leadership occurred during the implementation of the proactive conservation paradigm shift described above. The new leaders demonstrated support for the proactive conservation program, including additional cycles of competitive funding opportunities for proactive conservation projects. The fact that the proactive program persisted through transitions at multiple levels of leadership demonstrates progress made in proactive conservation becoming engrained in the culture of the Alaska USFWS office.

4. Discussion

Building a new program that is resilient to shifting priorities, declining budgets, and changes in internal leadership and administrations takes a significant investment in funds, staff, and time. In addition, it is important to look ahead to identify characteristics of a fully successful proactive conservation program so that movement toward a desired future state can be tracked and adaptively managed. The future vision of the proactive conservation program will be fully realized when thinking proactively is as natural and commonplace as reactive work is currently, and when staff are empowered and fully engaged working toward proactive priorities. A successful program is one where everyone shares a common understanding of proactive conservation goals. We anticipate that fully engaged employees will have increased productivity, work satisfaction, and retention, and support and promote a workplace that prioritizes learning and innovation. In our future vision, staff will place a high priority on establishing and maintaining strong relationships with trusted conservation partners within the context of advancing proactive conservation work.

Full engagement from leadership can help ensure that the proactive conservation program continues to focus on identified priorities. Proactive work is most successful when implemented on a landscape level and across disciplines. For the proactive conservation program to be successful, there must be clear and consistent messaging and support from leadership to staff. It should naturally be part of any conversation about program priorities and influence both current and future work. In addition, fostering and maintaining a collaborative working environment is a key component to a fully integrated program. When respect and trust are present, leadership and staff are more comfortable taking smart risks. Respectful and inclusive workplaces support innovation and innovation is key to the future of conservation.

4.1 Lessons learned in addressing barriers to implementation

As previously discussed, the pace and scale of climate change effects in Alaska prompted a re-examination of the conservation strategy used by the Alaska USFWS Office. A paradigm shift was required to institutionalize this new approach. However, implementing a major shift in thinking at the workplace is not without issues. There were a myriad of obstacles and challenges encountered in this process. The pace and magnitude of the shift elicited strong emotions from some staff, including the fear of change; an increase in uncertainty; the need for a clear vision for the new program and paradigm; time, capacity and skill set constraints for individuals; and a need for clear communication and collaboration. Addressing these issues was critical to the adoption of the shift toward proactive conservation.

4.1.1 Addressing the fear of change

There was an undercurrent of fear in many of the responses to surveys administered when the proactive initiative was first introduced to staff and management. Fear took many forms, including a fear of change, a fear of not having enough time to do proactive work, a fear of failing, and a fear of not having enough time to work on non-proactive projects. Working with staff to develop streamlining products to increase efficiency in conducting their traditional work helped create space for proactive work. In addition, institutionalizing the balance between traditional reactionary work and proactive work in staff workplans, and reflecting this balance in annual evaluations, provided a clear message that there was a place for both types of work in the workday, and established accountability for staff and management. While initially it was felt that providing staff with an opportunity to take a risk and try a different approach was enough, we learned that it was necessary to more actively direct staff to participate in the paradigm shift and to support their risk-taking by clearly acknowledging that some attempts at proactive conservation may fail, while others might succeed.

Recognizing that any change is difficult, leadership listened to staff discuss their fears and took action to minimize the risk associated with engaging in proactive conservation. Efforts focused on acknowledging and attempting to reduce fear by empowering staff with ownership over the path forward, adjusting the timing of implementation, maintaining open channels of communication, and utilizing a clear and transparent process. We learned we needed to be adaptive and allow the content and pace of the paradigm shift to be staff driven.

4.1.2 Addressing uncertainty

It was recognized that proactive conservation could mean a lot of different things to different people. A vision document was created in response to the desire of staff to have a clear statement of what program leaders meant by the shift to proactive conservation. At each stage of the paradigm shift, we were reminded how important it was for the messaging to be clear, consistent, and regularly conveyed by all levels of program leadership. Examples of proactive conservation projects were developed to address the concerns voiced by staff that they did not know how to implement a proactive project.

The staff in the program, like many conservation professionals, are so connected to their work that it often becomes a part of their identity. Therefore, it was unnerving and threatening when they are asked to step away from their traditional work or methods where they have demonstrated high competency, to now try something more experimental and less predictable in outcome. The primary metric for evaluating success in the past was through specific deliverables. To preserve the individual's sense of competence while pivoting toward more proactive work, we learned that a shift was required at all levels of the organization to measure value by conservation gains rather than products produced. Real progress will come when there are multiple successful and readily recalled examples of proactive conservation that are recognized by staff.

4.1.3 Addressing time, capacity, and skill set constraints

Staff were very concerned about not having time to engage in proactive work. While they voiced strong support and interest for proactive conservation, many also expressed concern that proactive work would require more time and energy than their normal work. This was of concern when they already felt time stressed. As a result, proactive work had the potential to be dismissed as an initiative that could only be pursued if their normal workload allowed them additional time. This staff input taught us the importance of streamlining current work to create space for employees to pursue proactive conservation work.

Another concern from staff was that they did not possess the correct skills to successfully complete proactive work and did not know if their training in traditional biology coursework prepared them to be successful at proactive conservation. Many of the respondents thought that they lacked skills in areas such as communication, facilitation, risk analysis, social science, public education, and outreach. To resolve this perceived deficiency, we sought opportunities to collaborate with practitioners who had those skills. Staff also identified the need for a leader to guide employees and a facilitator to help staff engage and be successful at proactive work. These needs were addressed by adding the two new positions within the proactive program mentioned previously.

4.1.4 Addressing the need for clear communication and collaboration

Communication internally and externally has consistently been identified as a priority to ensure a common understanding of what is meant by proactive conservation, identify opportunities for collaboration, and obtain the support of leadership and conservation partners. Previously, taking time to learn more about the work of others internally and externally was seen as a luxury rather than an essential awareness necessary to be able to identify opportunities for proactive work. To demonstrate the importance of these conversations, we created opportunities for more informal presentations and discussion about ongoing proactive projects. As a result of these efforts and others, there were improvements in awareness of the ongoing work in which colleagues were engaged, collaboration across offices, exchange of approaches for applied research and collaborative partnering, and sharing of work to meet important deadlines or to maximize conservation impacts.

5. Conclusions

Implementation of transformational initiatives is contingent on individuals and systems being willing to allow and embrace disruption. Challenges to the status quo can be threatening, with individuals and programs stressed by these changes. Real change takes leadership and perseverance of people working together for a common purpose. The proactive conservation approach employed by the Alaska USFWS office required risk, innovation, courage, and trust to act courageously to meet the conservation challenges of today and those posed by a rapidly changing future.

Both the magnitude and pace of the change required a bold, sustained, and adaptive approach to foster adoption of the paradigm shift and address stress and fear triggered by the proposed changes. Efforts to provide clarity and reduce uncertainty started with the creation of a vision document and continued with the issuance of an implementation strategy. The decision to solicit staff and management input, and to provide examples of on-the-ground proactive conservation was made in direct response to the concerns and requests from staff who wanted more direction on developing and implementing proactive projects.

The expectation that proactive work should constitute a percentage of staff and management workloads created the opportunity for employees to pursue proactive conservation work. In addition, two new proactive positions were created to facilitate and guide proactive efforts by engaging those with the needed skill sets, creating outreach and partnership opportunities, and connecting people interested in proactive work to pursue collaborative work. Benchmarks of success in the transformation toward an ingrained proactive conservation program in the Alaska USFWS office will include having multiple successful and readily recalled projects and efforts recognized as successful examples of proactive conservation, having most staff and management embrace and value the proactive approach, and having a program staffed with individuals capable of implementing proactive work as standard practice.

This paradigm shift is possible for all who are interested in adapting it to their own agencies, offices, and programs. It starts first with the desire to change the paradigm by which biologists approach their work by allowing biologists to work in anticipation of forthcoming threats alongside the traditional reactive approach. The process we present provides an experiment in attempting to change how people approach their conservation work in a federal workplace. We describe the issues that can arise and actions that can be taken during the shift toward proactive conservation, and we provide a framework for others to use to adapt to the needs of their offices. We highly encourage those facing pressing conservation needs to consider adapting their programs to better meet those challenges. We are hopeful that this framework will both inspire others to attempt a similar paradigm shift and that the lessons learned will serve to shorten the time in which those shifts can occur.

Acknowledgements

The authors would like to thank M. Daigneault, D. Wigglesworth, B. Meyer, D. Crane, P. Fasbender, S. Boario, and the staff and managers in the Alaska Regional Office and Northern and Southern Alaska Fish and Wildlife Conservation Offices who have supported the proactive concept. Thank you to N. Bjornlie, C. Burns, D. Crane, and R. Wilson for their early review of this manuscript. We would also like to thank the staff and managers in the USFWS Alaska Region for donating their time to be interviewed on the concept of proactive conservation. The findings and conclusions in this article are those of the authors and do not necessarily represent the views of the U.S. Fish and Wildlife Service.

Author details

Mary A. Colligan[1*†], Suzanne L. Nelson[2] and Melissa L. Burns[1,3]

1 U.S. Fish and Wildlife Service, Anchorage, Alaska, United States

2 U.S. Fish and Wildlife Service, Portland, Oregon, United States

3 U.S. Fish and Wildlife Service, Sacramento, CA, United States

*Address all correspondence to: macolligan@gmail.com

† Retired

References

[1] Donlan C. Proactive Strategies for Protecting Species: Prelisting Conservation and the Endangered Species Act. Oakland, California: The University of California Press; 2015

[2] Groom M, Meffe G, Carroll C. Principles of Conservation Biology. 3rd ed. Sunderland, Massachusetts: Sinauer Associates, Inc.; 2005

[3] Bruckner A, Hourigan T. Proactive management for conservation of *Acropora cervicornis* and *Acropora palmata*: Application of the US Endangered Species Act. In: Proceedings of the 9th International Coral Reef Symposium; 23-27, 2000; Bali, Indonesia. Oakland, California: University of California Press; 2000

[4] Fischer J, Hartel T, Kuemmerle T. Conservation policy in traditional farming landscapes. Conservation Letters. 2012;**5**:167-175

[5] Cardador L, Brotons L, Mougeot F, Giralt D, Bota G, Pomarol M, et al. Conservation traps and long-term species persistence in human-dominated systems. Conservation Letters. 2015;**8**:456-462

[6] Suzuki N, Parker K. Proactive conservation of high-value habitat for woodland caribou and grizzly bears in the boreal zone of British Columbia, Canada. Biological Conservation. 2019;**230**:91-103

[7] Shogren J, Tschirhart J, Anderson T, Whritenour Ando A, Beissinger S, Brookshire D, et al. Why economics matters for endangered species protection. Conservation Biology. 1999;**13**:1257-1261

[8] Drechsler M, Eppink F, Watzold F. Does proactive biodiversity conservation save costs? Biodiversity and Conservation. 2011;**20**:1045-1055

[9] Bush A, Hermoso V, Linke S, Nipperess D, Turak E, Hughes L. Freshwater conservation planning under climate change: Demonstrating proactive approaches for Australian Odonata. Journal of Applied Ecology. 2014;**51**:1273-1281

[10] Lader R, Walsh J, Bhatt U, Bieniek P. Projections of twenty-first-century climate extremes for Alaska via dynamical downscaling and quantile mapping. Journal of Applied Meteorology and Climatology. 2017;**56**(9):2393-2409. DOI: 10.1175/JAMC-D-16-0415.1

[11] Dunham K, Tucker A, Koons D, Abebe A, Dobson S, Grand J. Demographic responses to climate change in a threatened Arctic species. Ecology and Evolution. 2021;**11**(15):10627-10643. DOI: 10.1002/ece3.7873

[12] Marcot B, Jorgenson M, Lawler J, Hande C, DeGange A. Projected changes in wildlife habitats in Arctic natural areas of northwest Alaska. Climate Change. 2015;**130**:145-154

[13] Baltensperger A, Morton J, Huettmann F. Expansion of American marten (*Martes americana*) distribution in response to climate and landscape change on the Kenai Peninsula, Alaska. Journal of Mammalogy. 2017;**98**(3):703-714

[14] Magness D, Morton J. Using climate envelope models to identify potential ecological trajectories on

the Kenai Peninsula, Alaska. PLoS ONE. 2018;**13**:12. DOI: 10.137/journal.pone.0208883

[15] Abrahms B, DePietro D, Graffis A, Hollander A. Managing biodiversity under climate change: Challenges, frameworks, and tools for adaptation. Biodiversity and Conservation. 2017;**26**:2277-2293

[16] Magness D, Morton J, Huettmann F, Chapin F, McGuire A. A climate-change adaptation framework to reduce continental-scale vulnerability across conservation reserves. Ecosphere. 2011;**2**(10):1-23

[17] Williams J. RAD: A paradigm, shifting. Bioscience. 2021;**72**(1):13-15. DOI: 10.1093/biosci/biab123

[18] Magness D, Hoang L, Belote R, Brennan J, Carr W, Chapin F, et al. Management foundations for navigating ecological transformation by resisting, accepting, or directing social–ecological change. Bioscience. 2021;**71**:30-44. DOI: 10.1093/biosci/biab083

[19] Heifetz R, Grashow A, Linsky M. Leadership in a (permanent) crisis. Harvard Business Review. 2009;**87**:62-69. Available from: https://hbr.org/2009/07/leadership-in-a-permanent-crisis

[20] Heifetz R, Linsky M. A survival guide for leaders. Harvard Business Review. 2002;**80**:65-74. Available from: https://hbr.org/2002/06/a-survival-guide-for-leaders

Chapter 4

Endemic Fish Species in West Africa: Description, Spatial Distribution and Risk of Extinction Due to Global Change

Oumar Sadio

Abstract

West Africa is crossed by several major rivers, including Niger, Senegal, Volta, and Gambia. These rivers, along with their tributaries and deltas, and the coastal zone form a complex fluvial network that supports a variety of aquatic habitats and significant biodiversity. The aquatic ecosystems of West Africa are characterized by high biodiversity and strong connections to local livelihoods. In this study, a total of 112 fish, belonging to 27 families and 19 orders, have been recorded in West African countries. This includes one elasmobranch (Raja herwigi) and 111 actinopterygians. Guinea ranks first with 34% of the species, followed by Nigeria with 19%, and Cape Verde with 15%. Siluriformes dominate with 32% of the species, followed by Cypriniformes with 19%, and Cyprinodontiformes with 15%. Mochokidae has the highest number of endemic species, followed by Cyprinidae and Nothobranchidae. A total of 12 families each have only one endemic species. Benthopelagic species dominate, followed by demersal species. Freshwater species dominate with 83% of the total, marine species account for 18%, and one species (Limbochromis robertsi) inhabits both freshwater and brackish water. Most endemic species have low vulnerability to fishing while 12% are critically endangered. More than half are at risk. This situation calls for conservation policies.

Keywords: endemic species, global change, IUCN red list, risk of extinction, West Africa

1. Introduction

The Atlantic Ocean coastline of West Africa includes various marine environments such as coral reefs, seagrass beds, open ocean, and inland waters. Coastal environments, like estuaries, lagoons and bays are essential for the development of fishery resources due to their ecological and economic importance; they act as nurseries and feeding habitats for numerous commercial fish species, particularly during their early life stages [1, 2]. West African aquatic ecosystems (coastal zones, estuaries, rivers, lakes, and lagoons) are incredibly diverse and dynamic, shaped by

IntechOpen

the region's varied climate, geography, and hydrology. They are especially important for their high biodiversity and ecological functions [3]. These ecosystems are crucial habitats for marine life, including fish, crustaceans, and mollusks. These ecosystems support a wide range of fish species, ranging from the most cosmopolitan to endemics. Endemic fishes are an important part of the natural heritage of each country. Their conservation has implications on a worldwide basis since, by definition, an endemic taxon is found nowhere else. Areas with significant numbers of endemics and/or systematically significant endemics are prime candidates for conservation [4]. Increasing temperatures and changing rainfall patterns affect water levels and ecosystem health. Climate change, Overfishing, Industrial discharge, agricultural runoff, and urban waste contribute to water pollution, affecting aquatic life like fish species including endemic fish species. Species checklists play a fundamental role in ecosystem management, conservation planning, and environmental impact assessment by providing comprehensive insights into biodiversity and aiding in the efficient distribution of conservation resources [5]. This chapter aims to characterize the diversity of endemic fish species in West Africa and to analyze the risks of extinction as a result of climate change. Information regarding the endemic fish fauna is available in FishBase [6].

2. Methodology

The study area covers all marine, coastal, estuarine, fluvial and lacustrine aquatic ecosystems. The study covers 10 West African countries: Guinea, Nigeria, Cape Verde, Ivory Coast, Ghana, Liberia, Benin, Mali, Togo, and Senegal (**Figure 1**). All information on endemic species comes from FishBase (Froese and Pauly [6], https://fishbase.se/search.php) and stored in an Excel table. RStudio and Excel were used to plot the graphs.

3. Results

3.1 Breakdown of species by country

A total of 112 fish species belonging to 27 families and 19 orders were recorded in West African countries, comprising one elasmobranch (*Raja herwigi*, a Rajiforme belonging to Rajidae) and 111 actinopterygians (**Table 1**). Guinea ranks first with 34% of species (38 fish species), followed by Nigeria with 19% (n = 21), Cape Verde with 15% (n = 17), and Côte d'Ivoire with 11% (n = 12). Ghana (8%) and Liberia (7%) come fifth and sixth, respectively. Mali, Togo, and Benin each have 2% and Senegal 1% of endemic fish species (**Figure 2**).

3.2 Proportion des familles par ordre

In West Africa, the best-represented order of endemic species is Siluriformes with 36 families, followed by Cypriniformes with 19 families, Cyprinodontiformes with 17 families, and Cichliforme with 10 families. The remaining 15 orders are represented by fewer than 10 families (**Figure 3**). There are 10 orders represented by a single family (**Figure 3**).

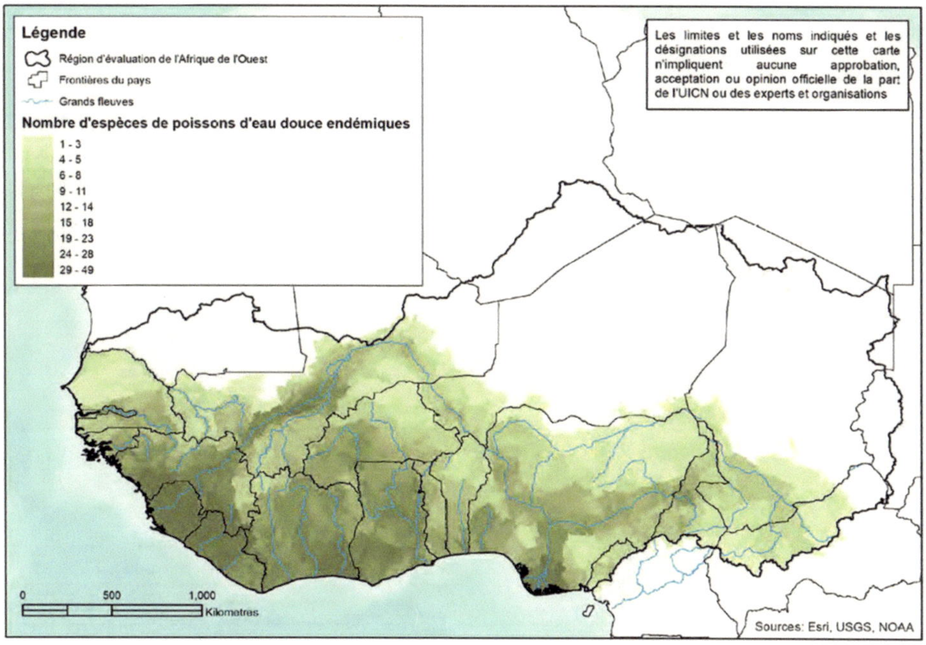

Figure 1.
Richness of regionally endemic freshwater fish species in West Africa based on Red List range maps. Source: Compiled by the authors of the report using IUCN Red List data (2021) (Source: Lalèyè and al. in Starnes and Darwall [3].

Pays	Order	Family	Species
Benin	Cypriniformes	Cyprinidae	*Barboides britzi*
Benin	Characiformes	Distichodontidae	*Nannocharax signifer*
Cap-Vert	Gobiesociformes	Gobiesocidae	*Apletodon barbatus*
Cap-Vert	Ophidiiformes	Ophidiidae	*Bassozetus oncerocephalus*
Cap-Vert	Mugiliformes	Mugilidae	*Chelon bispinosus*
Cap-Vert	Ovalentaria_misc	Pomacentridae	*Chromis lubbocki*
Cap-Vert	Eupercaria/misc	Sparidae	*Diplodus fasciatus*
Cap-Vert	Eupercaria/misc	Sparidae	*Diplodus lineatus*
Cap-Vert	Eupercaria/misc	Sparidae	*Diplodus prayensis*
Cap-Vert	Centrarchiformes	Girellidae	*Girella stuebeli*
Cap-Vert	Gobiiformes	Gobiidae	*Gobius ateriformis*
Cap-Vert	Gobiiformes	Gobiidae	*Gobius salamansa*
Cap-Vert	Gobiiformes	Gobiidae	*Gobius tetrophthalmus*
Cap-Vert	Blenniiformes	Blenniidae	*Microlipophrys caboverdensis*
Cap-Vert	Pleuronectiformes	Soleidae	*Pegusa cadenati*
Cap-Vert	Beloniformes	Belonidae	*Platybelone lovii*
Cap-Vert	Rajiformes	Rajidae	*Raja herwigi*
Cap-Vert	Ovalentaria_misc	Pomacentridae	*Similiparma hermani*

Pays	Order	Family	Species
Cap-Vert	Eupercaria/misc	Sparidae	*Virididentex acromegalus*
Ivory Coast	Characiformes	Alestidae	*Brycinus derhami*
Ivory Coast	Siluriformes	Mochokidae	*Chiloglanis normani*
Ivory Coast	Cichliformes	Cichlidae	*Chromidotilapia cavalliensis*
Ivory Coast	Siluriformes	Clariidae	*Clarias lamottei*
Ivory Coast	Cypriniformes	Cyprinidae	*Enteromius traorei*
Ivory Coast	Cyprinodontiformes	Nothobranchiidae	*Epiplatys etzeli*
Ivory Coast	Siluriformes	Malapteruridae	*Malapterurus thysi*
Ivory Coast	Osteoglossiformes	Mormyridae	*Marcusenius elegans*
Ivory Coast	Osteoglossiformes	Mormyridae	*Pollimyrus eburneensis*
Ivory Coast	Siluriformes	Mochokidae	*Synodontis comoensis*
Ivory Coast	Siluriformes	Mochokidae	*Synodontis koensis*
Ivory Coast	Siluriformes	Mochokidae	*Synodontis punctifer*
Ghana	Siluriformes	Claroteidae	*Chrysichthys walkeri*
Ghana	Cypriniformes	Cyprinidae	*Enteromius guildi*
Ghana	Cypriniformes	Cyprinidae	*Enteromius subinensis*
Ghana	Cypriniformes	Cyprinidae	*Enteromius walkeri*
Ghana	Siluriformes	Schilbeidae	*Irvineia voltae*
Ghana	Cichliformes	Cichlidae	*Limbochromis robertsi*
Ghana	Siluriformes	Malapteruridae	*Malapterurus tanoensis*
Ghana	Cyprinodontiformes	Nothobranchiidae	*Pronothobranchius seymouri*
Ghana	Siluriformes	Mochokidae	*Synodontis macrophthalmus*
Guinea	Siluriformes	Amphiliidae	*Amphilius kakrimensis*
Guinea	Characiformes	Alestidae	*Brycinus carolinae*
Guinea	Cyprinodontiformes	Nothobranchiidae	*Callopanchax sidibeorum*
Guinea	Siluriformes	Mochokidae	*Chiloglanis camarabounyi*
Guinea	Siluriformes	Mochokidae	*Chiloglanis dialloi*
Guinea	Siluriformes	Mochokidae	*Chiloglanis kabaensis*
Guinea	Siluriformes	Mochokidae	*Chiloglanis kolente*
Guinea	Siluriformes	Mochokidae	*Chiloglanis lamottei*
Guinea	Siluriformes	Mochokidae	*Chiloglanis loffabrevum*
Guinea	Siluriformes	Mochokidae	*Chiloglanis longibarbis*
Guinea	Siluriformes	Mochokidae	*Chiloglanis nzerekore*
Guinea	Siluriformes	Mochokidae	*Chiloglanis pezoldi*
Guinea	Siluriformes	Claroteidae	*Chrysichthys levequei*
Guinea	Cichliformes	Cichlidae	*Coptodon konkourensis*
Guinea	Cichliformes	Cichlidae	*Coptodon rheophila*
Guinea	Cypriniformes	Cyprinidae	*Enteromius anniae*

Pays	Order	Family	Species
Guinea	Cypriniformes	Cyprinidae	*Enteromius cadenati*
Guinea	Cypriniformes	Cyprinidae	*Enteromius ditinensis*
Guinea	Cypriniformes	Cyprinidae	*Enteromius guineensis*
Guinea	Cypriniformes	Cyprinidae	*Enteromius raimbaulti*
Guinea	Cypriniformes	Cyprinidae	*Enteromius teugelsi*
Guinea	Cyprinodontiformes	Nothobranchiidae	*Epiplatys guineensis*
Guinea	Cypriniformes	Cyprinidae	*Labeo rouaneti*
Guinea	Cypriniformes	Danionidae	*Leptocypris konkoureensis*
Guinea	Siluriformes	Malapteruridae	*Malapterurus teugelsi*
Guinea	Synbranchiformes	Mastacembelidae	*Mastacembelus kakrimensis*
Guinea	Siluriformes	Mochokidae	*Microsynodontis polli*
Guinea	Cyprinodontiformes	Nothobranchiidae	*Nimbapanchax melanopterygius*
Guinea	Siluriformes	Amphiliidae	*Paramphilius teugelsi*
Guinea	Cypriniformes	Danionidae	*Raiamas levequei*
Guinea	Cyprinodontiformes	Procatopodidae	*Rhexipanchax kabae*
Guinea	Cyprinodontiformes	Procatopodidae	*Rhexipanchax lamberti*
Guinea	Cyprinodontiformes	Nothobranchiidae	*Scriptaphyosemion nigrifluvi*
Guinea	Siluriformes	Mochokidae	*Synodontis dekimpei*
Guinea	Siluriformes	Mochokidae	*Synodontis levequei*
Guinea	Siluriformes	Mochokidae	*Synodontis tourei*
Guinea	Cichliformes	Cichlidae	*Wallaceochromis rubrolabiatus*
Guinea	Cichliformes	Cichlidae	*Wallaceochromis signatus*
Liberia	Cichliformes	Cichlidae	*Coptodon coffea*
Liberia	Cypriniformes	Cyprinidae	*Enteromius boboi*
Liberia	Cypriniformes	Cyprinidae	*Enteromius carcharhinoides*
Liberia	Cypriniformes	Cyprinidae	*Enteromius melanotaenia*
Liberia	Cypriniformes	Cyprinidae	*Enteromius trispiloides*
Liberia	Cyprinodontiformes	Nothobranchiidae	*Epiplatys cashneri*
Liberia	Siluriformes	Amphiliidae	*Paramphilius firestonei*
Liberia	Cyprinodontiformes	Nothobranchiidae	*Scriptaphyosemion schmitti*
Mali	Cyprinodontiformes	Procatopodidae	*Micropanchax ehrichi*
Mali	Siluriformes	Mochokidae	*Synodontis gobroni*
Nigeria	Characiformes	Alestidae	*Alestopetersius smykalai*
Nigeria	Characiformes	Alestidae	*Arnoldichthys spilopterus*
Nigeria	Anabantiformes	Anabantidae	*Ctenopoma nebulosum*
Nigeria	Siluriformes	Amphiliidae	*Doumea reidi*
Nigeria	Cypriniformes	Cyprinidae	*Enteromius clauseni*
Nigeria	Cyprinodontiformes	Nothobranchiidae	*Fundulopanchax arnoldi*
Nigeria	Cyprinodontiformes	Nothobranchiidae	*Fundulopanchax clauseni*

Pays	Order	Family	Species
Nigeria	Cyprinodontiformes	Nothobranchiidae	*Fundulopanchax deltaensis*
Nigeria	Cyprinodontiformes	Nothobranchiidae	*Fundulopanchax powelli*
Nigeria	Osteoglossiformes	Mormyridae	*Marcusenius kainjii*
Nigeria	Characiformes	Distichodontidae	*Neolebias axelrodi*
Nigeria	Characiformes	Distichodontidae	*Neolebias powelli*
Nigeria	Siluriformes	Claroteidae	*Notoglanidium akiri*
Nigeria	Siluriformes	Claroteidae	*Parauchenoglanis buettikoferi*
Nigeria	Cichliformes	Cichlidae	*Pelvicachromis sacrimontis*
Nigeria	Cichliformes	Cichlidae	*Pelvicachromis silviae*
Nigeria	Siluriformes	Mochokidae	*Synodontis guttatus*
Nigeria	Siluriformes	Mochokidae	*Synodontis omias*
Nigeria	Siluriformes	Mochokidae	*Synodontis robbianus*
Nigeria	Siluriformes	Mochokidae	*Synodontis xiphias*
Nigeria	Cichliformes	Cichlidae	*Thysochromis annectens*
Senegal	Gobiiformes	Gobiidae	*Corcyrogobius pulcher*
Togo	Cyprinodontiformes	Procatopodidae	*Micropanchax bracheti*
Togo	Cyprinodontiformes	Procatopodidae	*Micropanchax keilhacki*

Table 1.
List of the 112 endemic fish species with country, order and family.

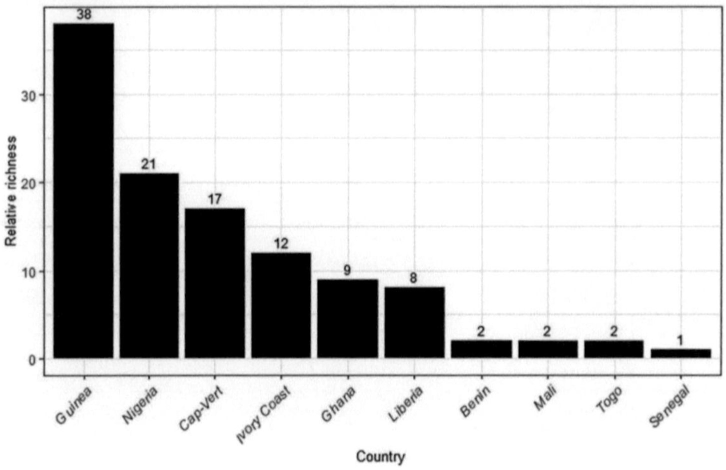

Figure 2.
Number of endemic fish species by country of West African.

3.3 Proportion des espèces par famille de poissons

Of the 27 families grouping endemic species in West Africa, the best represented is the Mochokidae family with 23 endemic fish species, followed by the Cyprinidae family with 17 species, the Nothobranchiidae family with 12 species, and the Cichidae family with 10 species (**Figure 4**). The remaining 23 families are represented by fewer than 10 species. There are 12 families that are represented by only one endemic species in West Africa (**Figure 4**).

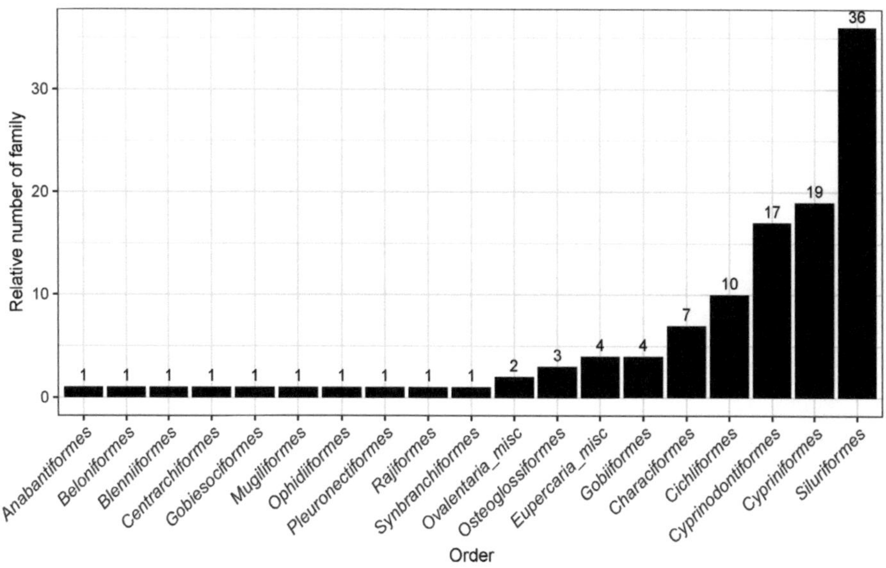

Figure 3.
Number of families per orders recorded from 10 West African countries.

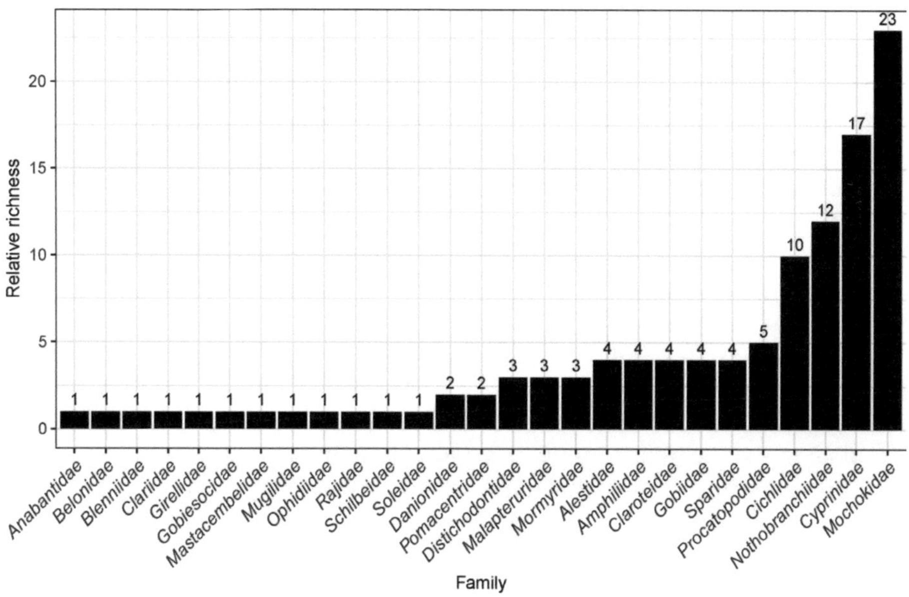

Figure 4.
Number of species per families recorded from 10 West African countries.

3.4 Habitat guild

Analysis of the habitat guide revealed five groups of endemic fish species in West Africa. Benthic pelagics are the most diverse, with 57 species representing 51% of the endemic species in West Africa, followed by demersal (38%, 38 species), pelagic (18%, 14 species), reef and bathydemersal with 2% and 1%, respectively (**Figure 5**).

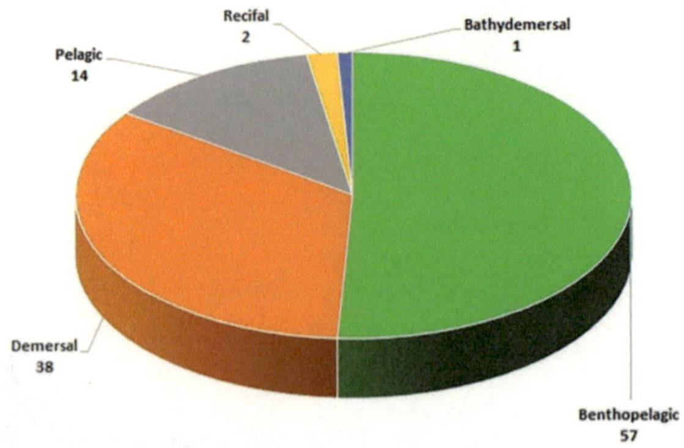

Figure 5.
Distribution of species richness by habitat guild for 10 West African countries.

3.5 Ecological guild

The 112 endemic species from West Africa are distributed between 3 ecological guides, namely Freshwater with 93 species, marine with 18 species, and both freshwater and brackish water with only 1 species (**Figure 5**). One fish species (*Limbochromis robertsi*) frequents both freshwater and brackish water.

3.6 Guilde trophique

The distribution of species according to their trophic guild has made it possible to identify 103 omnivorous species (92%), 5 carnivorous species (4%), 3 herbivore species (3%), and one detritivore (1%) in West African aquatic ecosystems (**Figure 6**).

3.7 Vulnerability to fishing

The analysis of vulnerability to fishing revealed that the 112 endemic West African species are grouped into four different groups. There are 103 species in the low vulnerability group, representing 91% of the total richness, compared with four species in the moderate vulnerability group, three species in the low to moderate vulnerability group, and two species in the moderate to high vulnerability group.

3.8 Statut IUCN

The 112 species endemic to West Africa are grouped into seven different groups on the IUCN Red List (Not Evaluated, Data deficient, Least concern, Vulnerable, Near Threatened, Endangered, Critically endangered). Analysis of the IUCN status reveals the dominance of Endangered species, followed by Least concern, Not Evaluated, Vulnerable and Critically endangered. Of the 112 endemic species listed, half are threatened (Vulnerable, Endangered, Critically endangered). Species in the Least Concern group are well represented among the endemic species in West Africa. The Near threatened group is the least represented (**Figure 7**). Considering the IUCN assessments at the global level 24% of the species are classified as "Endangerd" with 27 species, 21% as " Least concern, n = 23", 20% as " Not Evaluated, n = 22", 18% as

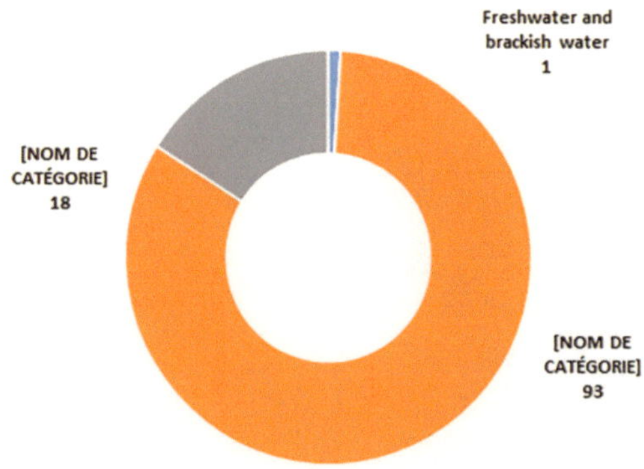

Figure 6.
Distribution of species richness by ecological guild for 10 West African countries.

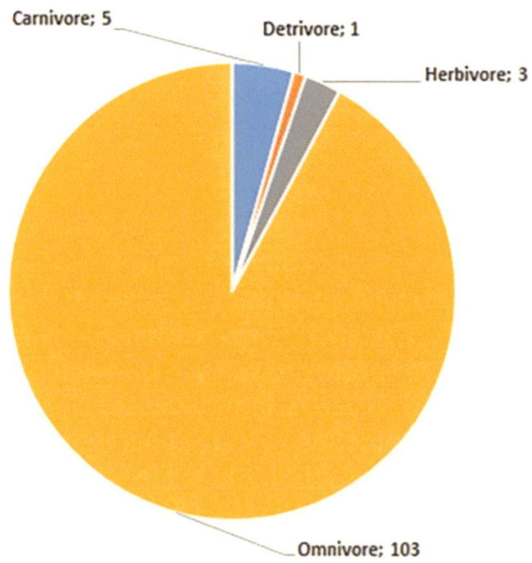

Figure 7.
Distribution of species richness by trophic guild for 10 West African countries.

"Vulnerable, n = 20", 12% as "Criticaly endangered, n = 13", 4% as "Data deficient, n = 5", and 2% as "Near threatened, n = 2" (**Figure 7**).

3.9 Global extinction risk of endemic fish species

The families with species in the Critical Endangered category are Mochokidaee (5 species), Cyprinidae (4 species), Nothobranchiidae (2 species), Chiclidae (1 species), and Claroteidae (1 species) (**Figure 8**). The Endangered status concerns Mochokidae, Alestidae, and Nothobranchiidae, with 8, 4, and 3 species, respectively. Cyprinidae, Chiclidae, Claroteidae, and Distichodontidae each have two species in the Endangered category, while Amphiliidae, Anabantidae, Procatopodidae, and Schilbeidae each have only one species in the Endangered category. Malapteruridae and Nothobranchiidae are the two families with one species each in the Near Threatened category (**Figure 8**). The families with species in the Vulnerable category are Cyprinidae (six species), Mochokidae (three species), Nothobranchiidae

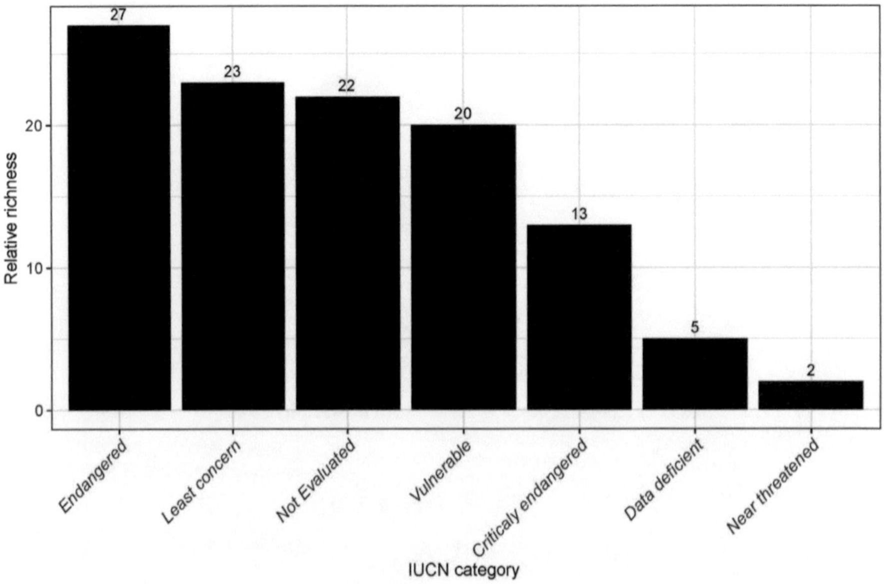

Figure 8.
Distribution of endemic fish species by IUCN category for 10 West African countries.

and Danionidae with two species each, and Danionidae, Gobiidae, Amphiliidae, Malapteruridae, Procatopodidae, Claroteidae, Clariidae and Gobiesocidae with a single species each (**Figure 8**). There are 23 species belonging to 12 different families in the Least Concern category, 5 species in the Data Deficient category in 5 different families, and 22 species in 12 different families in the Not Evaluated category (**Figure 8**). Different levels of extinction risk are apparent at family level. The families concerned by Critical Endangered are the Mokokidae (38% of 112 species), followed by Cyprinidae (31%), Nothobranchiidae (15%), Claroteidae (8%), and Cichlidae (8%). More specifically, these are the species *Synodontis dekimpei, Synodontis guttatus, Synodontis macrophthalmus, Synodontis tourei,* and *Synodontis xiphias* in the Mochokidae family. In the Cyprinidae family, the Critically Endangered species are *Enteromius boboi, Enteromius carcharhinoides, Enteromius clauseni,* and *Enteromius melanotaenia*. *Fundulopanchax powelli* and *Scriptaphyosemion schmitti* are the two Critically Endangered species of the family Nothobranchiidae. *Coptodon coffea* and *Parauchenoglanis buettikoferi*, belonging to the families Cichlidae and Claroteidae respectively, are in the Critical Endangered category (**Figure 9**). With regard to the Endangered Category, the families concerned are the Mokokidae (30% of species) with the following species: *Chiloglanis dialloi, Chiloglanis kabaensis, Chiloglanis kolente, Chiloglanis loffabrevum, Chiloglanis longibarbis, Chiloglanis normani, Chiloglanis nzerekore* and *Chiloglanis pezoldi*. The Alestidae comes second with 15% of species Endangered, and is represented *Alestopetersius smykalai, Arnoldichthys spilopterus, Brycinus carolinae*, and *Brycinus derhami*, all of which are endangered (**Figure 9**). The families with vulnerable status are Cyprinidae (30% of 112 species), Mochokidae (15%), Danionidae (10%), Nothobranchiidae (10%), Gobiesocidae, Gobiidae, Malapteruridae, Clariidae, Claroteidae, Aphiliidae and Procatopodidae, each with 5% of vulnerable species. The species in the Near Threatened category are *Malapterurus thysi* of the family Malapteruridae (50% of species) and *Fundulopanchax deltaensis* of the family Nothobranchiidae (50% of species) (**Figure 8**). The remaining

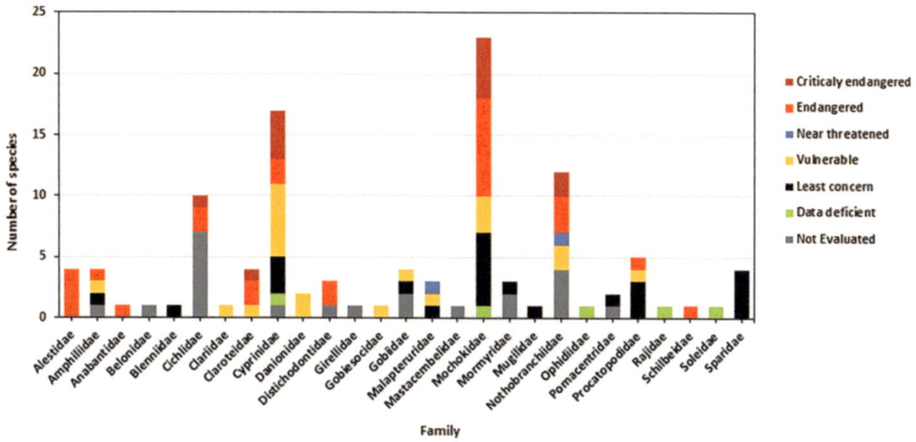

Figure 9.
Global extinction risk of endemic fish species in West Africa aquatic water.

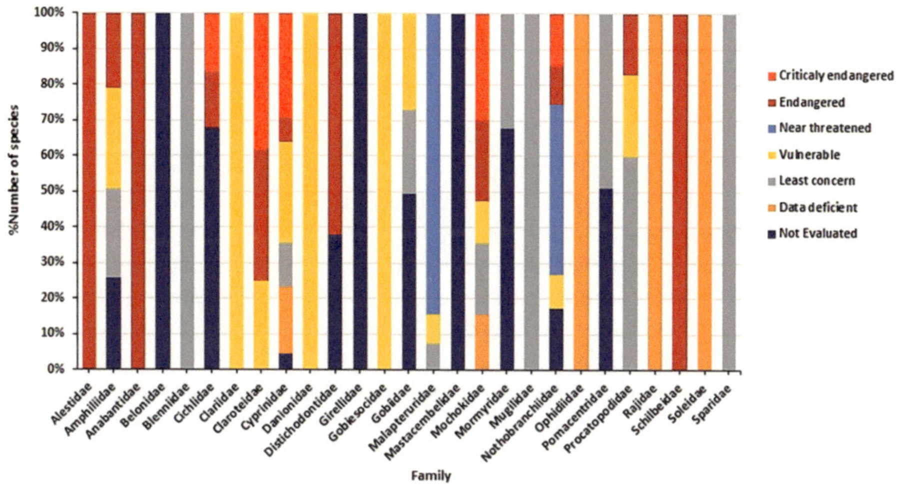

Figure 10.
Extinction risk in the various endemic fish species, expressed as a percentage of species richness per family.

87 species are classified as either Near Threatened, Least Concern, Not Evaluated, Data Deficient or Least Concern (**Figures 9** and **10**).

4. Discussion

This study, based on FishBase data, made it possible to list 112 endemic fish species spread across 10 West African countries [5]. They are much more common in Guinea, Nigeria, Cape Verde, and the Ivory Coast. Endemic fish species are found mainly in rivers but also in coastal areas. This reflects good diversity and wide geographical distribution [7–9]. The four best-represented orders are Siluriformes, Cypriniformes, Cyprinodontiformes and Cichliformes. At the family level, Mochokidae, Cyprinidae, Nothobranchiidae and Cichlidae are the four best-represented families. There are 12 monospecific families. The majority are omnivorous, freshwater, benthopelagic, and

demersal species, most of which are not targeted by fishing [10]. The 112 endemic species identified are divided into seven categories on the IUCN red list [11].
The presence of endemic species has been reported by several authors, including [12] in Greece and [13] in Iran. There are many more species in danger according to IUCN criteria [10]. Habitat loss [14, 15], climate change [16, 17] and overexploitation [18] are the main threats to endemic species in West Africa. Species in the Alestidae family (*Alestopetersius smykalai, Arnoldichthys spilopterus, Brycinus carolinae,* and *Brycinus derhami*), the Anabantidae family (*Ctenopoma nebulosum*), and the Schilbeidae family (*Irvineia voltae*) are the most threatened with extinction. The presence of endemic species in critical danger of extinction has been confirmed by several authors. There are 13 endemic species in critical danger of extinction due to their narrow distribution area and small population [19]. Some species with restricted distribution require greater conservation efforts, and some rarer species could be farmed in their natural environment to maintain their population. This would ensure their survival, provided that the capture of wild specimens is carefully controlled. Conservation issues are therefore necessary to preserve the endemic species of West Africa, as pointed out by [20]. Species that experience any of the above attributes must be given priority, monitored, and managed carefully in an effort to promote genetic conservation (**Figure 10**).

5. Conclusion

A total of 112 fish species belonging to 27 families and 19 orders were recorded in West African countries, comprising one elasmobranch (*Raja herwigi*, a Rajiforme belonging to Rajidae) and 111 actinopterygians. Guinea ranks first with 34% of species (38 fish species), followed by Nigeria with 19% (n = 21), Cape Verde with 15% (n = 17), and Côte d'Ivoire with 11% (n = 12). The best-represented order is Siluriformes, with 36 families, followed by Cypriniformes, with 19 families, Cyprinodontiformes, with 17 families, and Cichliforme, with 10 families. Of the 27 families grouping endemic species in West Africa, the best represented is the Mochokidae with 23 species, followed by the Cyprinidae with 17 species, Nothobranchiidae with 12 species, and Cichidae with 10 species. Analysis of the habitat guide revealed 5 groups (Benthic pelagics, 57 species; demersal, 38 species; pelagic, 14 species; reef, 2; and bathydemersal, 1 species). The 112 endemic species from West Africa are distributed between three ecological guides, namely Marine, Freshwater, and one fish species (*Limbochromis robertsi*) frequents both freshwater and brackish water. Most of them are omnivores, low vulnerability, endangered species. Of the 112 endemic species listed, half are threatened (Vulnerable, Endangered, and Critically endangered). Analyzing extinction risk, Mochokidae, Cyprinidae, Nothobranchiidae, Chiclidae, and Claroteidae are in critical endangered category. More specifically, these are the species *Synodontis dekimpei, Synodontis guttatus, Synodontis macrophthalmus, Synodontis tourei,* and *Synodontis xiphias* in the Mochokidae family. In the Cyprinidae family, the Critically Endangered species are *Enteromius boboi, Enteromius carcharhinoides, Enteromius clauseni*, and *Enteromius melanotaenia*. *Fundulopanchax powelli* and *Scriptaphyosemion schmitti* are the two Critically Endangered species of the family Nothobranchiidae. *Coptodon coffea* and *Parauchenoglanis buettikoferi*, belonging to the families Cichlidae and Claroteidae, respectively, are in the Critical Endangered category. These results show that there is a real risk of extinction of certain endemic species in West Africa, probably due to the impact of climate change rather than pollution or fishing.

Acknowledgements

The data processed here comes from the FishBase database. The author would like to warmly thank the FishBase team, headed by Raina Froese and Daniel Pauly, for making this important fish database available to the international scientific community.

Conflict of interest

The authors declare no conflict of interest.

Acronym and abbreviation

IUCN	International Union for Conservation of Nature

Author details

Oumar Sadio
French National Research Institute for Sustainable Development, Dakar, Senegal

*Address all correspondence to: oumar.sadio@ird.fr

References

[1] Elliott M, Whitfield AK, Potter IC, Blaber SJM, Cyrus DP, Nordlie FG, et al. The guild approach to 251 categorizing estuarine fish assemblages: A global review. Fisheries. 2007;**8**:241-268. DOI: 10.1111/j.1467-252 2679.2007.00253.x

[2] de Carvalho-Souza GF, González-Ortegón E, Baldó F, Vilas C, Drake P, Llope M. Natural and anthropogenic effects on 254 the early life stages of European anchovy in one of its essential fish habitats, the Guadalquivir estuary. Marine Ecology Progress Series. 2019;**255**(617-618):67-79. DOI: 10.3354/meps12562

[3] Starnes T, Darwall WRT. Identification et validation des Zones clés pour la biodiversité des eaux douces d'Afrique occidentale. Gland, Suisse: UICN; 2021

[4] Coad BW. Endemicity in the freshwater fishes of Iran. Iranian Journal of Animal Biosystematics. 2006;**1**:1-13

[5] Lo BS. Zoological checklists: From natural history museums to ecosystems. Diversity. 2023;**15**:741

[6] Froese R, Pauly D. FishBase. World Wide Web electronic publication. www.Fishbase.org.version (02/2024); 2024

[7] Lévêque C. La biodiversité au quotidien. Le développement durable à l'épreuve des faits. Versailles, Quæ/Paris: iRD Éditions; 2008

[8] Nelson JS. Fishes of the World. 5th ed. Hoboken, New Jersey, USA: John Wiley & Sons, Inc.; 2016

[9] Eschmeyer WN, Fong JD. Catalog of Fishes. California Academy of Sciences; 2024. Available from: http://research.calacademy.org/research/ichthyology/catalog/fishcatmain.asp/

[10] Daget J, Gosse J, Thys van den Audenaerde DFE. Poissons d'eau douce de l'Afrique de l'Ouest. Paris, France: ORSTOM; 1984

[11] IUCN. The IUCN Red List of Threatened Species. Retrieved from IUCN Red List. Pour des données actuelles sur le statut de conservation des espèces. 2024

[12] Barbieri RS, Zogaris E, Kalogianni MT, Stoumboudi Y, Chatzinikolaou S, Giakoumi Y, et al. Freshwater fishes and lampreys of Greece: An annotated checklist. In: Monographs on Marine Sciences No.8 Hellenic Centre for Marine Research. 130. Anavyssos, Greece: Hellenic Centre for Marine Research; 2015

[13] Esmaeili HR, Sayyadzadeh G, Özulug M, Geiger M, Freyhof J. Three new species of Turcinoemacheilus from Iran and Turkey (Teleostei: Nemacheilidae). Ichthyological Exploration of Freshwaters. 2014;**24**:257-273

[14] Dudgeon D, Arthington AH, Gessner MO, Kawabata ZI, Knowler DJ, Lévêque C, et al. Freshwater biodiversity: Importance, threats, status and conservation challenges. Biological Reviews. 2006;**81**:163-182

[15] Vörösmarty CJ, McIntyre PB, Gessner MO, Dudgeon D, Prusevich A, Green P, et al. Global threats to human water security and river biodiversity. Nature. 2010;**467**:555-561. DOI: 10.1038/nature09440

[16] Palmer MA, Lettenmaier DP, Poff NL, Sandra L, Postel SL,

Richter B, et al. Climate change and river ecosystems: Protection and adaptation options. Environmental Management. 2009;**44**:1053-1068. DOI: 10.1007/s00267-009-9329-1

[17] Yates D, Millington A. Impacts of climate change on freshwater ecosystems: Insights from observational data and modeling. Hydrological Sciences Journal. 2018;**63**:808-825. DOI: 10.1080/02626667.2018.1484722

[18] Pauly D, Christensen V, Guénette S, Pitcher TJ, Sumaila UR, Walters CJ, et al. Towards sustainability in world fisheries. Nature. 2002;**418**(6898):689-695. DOI: 10.1038/nature01017

[19] Işik K. Rare and endemic species: Why are they prone to extinction? Turkish Journal of Botany. 2011;**35**:411-417

[20] Barlow RG, Voss GL. The role of fish in West African ecosystems. Journal of Fish Biology. 2022;**100**:890-912

Chapter 5

Endemic Species of Elasmobranchs (Sharks and Rays) in South America: A Review

Ilka Branco-Nunes, Camila Araújo, Emmanuely Ferreira, Sibele Mendonça, Natalia Alves, Diogo Nunes, Danielle Viana and Paulo Oliveira

Abstract

Elasmobranchs (sharks and rays) generally have intrinsic biological characteristics (e.g., low fecundity, large body sizes, slow growth rates, late sexual maturity, and long lifespans) that make them highly vulnerable to anthropic actions. Predatory fishing, pollution, and degradation of natural habitats have been declining strongly the populations of sharks and rays worldwide. It is essential to understand the diversity of species that occur in this very limited group (~1200 species), in addition to identifying their level of endemism in coastal and island areas. In this context, several non-lethal techniques can be developed and used to help better understand the ecology and behavior of these sharks and rays, with the hope of contributing to the conservation of this fascinating group.

Keywords: diversity, ecology, conservation, non-lethal methods, marine protected area

1. Introduction

Living fishes are comprised of two primary taxonomic groups, which are classified as bony fishes (class Osteichthyes) and cartilaginous fishes (class Chondrichthyes). The class Osteichthyes makes up about 95% of modern fishes, including an impressive array of body forms. In general terms, the Chondrichthyans have a long evolutionary history, reaching back over 450 million years. Modern chondrichthyans are normally separated into two groups, the Elasmobranchii and the Holocephali [1]. The subclass Elasmobranchii is composed of sharks and batoids (skates and rays) with more than 1200 species across 70 families. The Holocephali is represented only by the chimeras, which include more than 50 species across three families, living mainly in the deep sea [2].

Despite the reduced number of species, compared to bony fish (~34,000 species), the elasmobranchs can inhabit a wide variety of environments. The sharks and batoids are essentially found in marine ecosystems, being present in all the world's oceans of both hemispheres, from the Arctic to the Antarctic. However, they also could occur in estuaries and freshwater habitats (e.g., stingrays from the Potamotrygonidae family). This diversity of habitats in which the group can occur is accompanied by a variety of sizes and shapes, ranging from classic aerodynamic-shaped sharks to dorsally flattened skates and rays [3]. Sharks and rays vary greatly in size, from individuals with a maximum total length of ~23 cm (e.g., the Pygmy ribbontail catshark, *Eridacnis radcliffei*) to the Whale shark, *Rhincodon typus*, the largest living fish species, which can reach up to ~17 m in length [4].

Although historically sharks and rays have been considered long-lived and slow-growing, there is a diverse range of growth patterns and life-history strategies found in chondrichthyans [5]. The differences in energetic absorption, use, and excretion of an individual, enable the most growth functions. For example, when a specimen reaches sexual maturity, there are indicators of increased energy designation to the reproductive system, thus transferring resources away from somatic growth and into reproductive growth. The differences in growth, between males and females, have already been extensively reported, with a pattern that females frequently get a larger size than males [6, 7].

All elasmobranchs exhibit internal fertilization, low fecundity, and late sexual maturity; however, there is an ample range of reproductive models displayed by the group. Particularly for females, it is possible to verify a variety of strategies to provision developing embryos with nutrients. In this sense, organisms can be classified as lecithotrophic, when the embryo's nutrition occurs exclusively through the yolk sac's reserves, and matrotrophic, for species where the female supplies embryos with yolk and one or more auxiliary sources of nutrition. Reproductive models are further classified about the site of embryonic development, with species designated as oviparous (e.g., nurse shark, *Ginglymostoma cirratum*), through the egg-laying in the environment, and viviparous, where embryonic development occurs entirely inside the female's uterus (e.g., pelagic stingray, *Pteroplatytrygon violacea*). In general lines, matrotrophic provisioning strategies further vary, making sharks and rays a single and diverse taxon [7].

The elasmobranchs occur at a variety of trophic levels, using a series of strategies to find and capture their prey, ranging from planktivorous species (e.g., Mobula rays) up to carnivorous predators (e.g., Tiger shark, *Galeocerdo cuvier*). The various strategies, from predators to mesopredators, that species adopt have a strong influence on their ecological roles in aquatic ecosystems [8]. In this sense, the habitat selection carried out by an animal is generally related to the food supply in the environment and consequent energy return, in addition to the choice of environments suitable for reproductive activities, such as the place of copulation and birth.

The intrinsic biological characteristics (e.g., low fecundity, large body sizes, slow growth rates, late sexual maturity, and long lifespans) reported here demonstrate how sharks and rays are highly vulnerable to anthropogenic actions. The identification of essential habitats for feeding and reproduction of elasmobranchs is especially important to preserve the balance and conservation of populations. Predatory fishing, pollution, and degradation of natural habitats have been declining strongly the populations of sharks and rays worldwide. In this sense, the present chapter aimed, in addition to reporting important biological characteristics mentioned here, to highlight the degree of diversity and endemism of elasmobranchs found in South America, in addition to simultaneously highlighting non-lethal alternatives for the development of sustainable research, used to help better understand the ecology and behavior of these sharks and rays, with the hope of contributing to the conservation of this fascinating group.

2. Materials and methods

2.1 Diversity, endemism, and species conservation status

Endemic species are found exclusively in a specific geographic region and are not found naturally in other parts of the world. The methodology used, in this chapter, to evaluate the diversity and endemism of elasmobranchs in South America was conducted through in-depth research carried out on the International Union for Conservation of Nature (IUCN)'s Red List of Threatened Species platform, which allowed the use of series of filters (e.g., habitat occupied by the species, taxonomic description, and area of occurrence, among others). Additional research on manuscripts was also carried out. From the data obtained, it was possible to reach a broad overview of the endemic species that occur in coastal and island environments, in addition to an assessment of their population status. A global total of 1194 elasmobranchs evaluated by the IUCN's Red List of Threatened Species was verified; among these, 256 correspond to sharks and rays that occur around South America, of which 38 (~15%) are endemic species. Among the marine elasmobranch's endemic to South America, 68.4% correspond to rays, and 31.6% were represented by sharks. The species were distributed into 16 families, with a greater number of families being found for batoids than for sharks, with the Rajidae family representing the largest number of skates (**Table 1**).

Scientific name	Family	Habitat	Geography rangy	IUCN
Atlantoraja castelnaui	Arhynchobatidae	MN, MDB	Brazil to Argentina	CR
Dipturus lamillai	Rajidae	MN, MDB	Falkland Islands*	LC
Discopyge castelloi	Narcinidae	MN	Argentina	DD
Dipturus chilensis	Rajidae	MN, MDB	Chile	EN
Psammobatis scobina	Arhynchobatidae	MN	Chile (Malvinas)*	LC
Etmopterus litvinovi	Emopteridae	MDB	Chile	LC
Rajella eisenhardti	Rajidae	MDB	Ecuador (Galápagos)*	LC
Urobatis marmoratus	Urotrygonidae	Unknown	Chile	DD
Rajella sadowskii	Rajidae	MDB	Brazil	LC
Benthobatis kreffti	Narcinidae	MDB	Brazil	VU
Bythaelurus giddingsi	Pentanchidae	MDB	Ecuador (Galápagos)*	LC
Centroscyllium granulatum	Etmopteridae	MDB	Chile	VU
Centroscymnus macracanthus	Somniosidae	MDB	Chile (Magellanes)	DD
Dipturus ecuadoriensis	Rajidae	Unknown	Ecuador	DD
Dipturus mennii	Rajidae	MN, MDB	Brazil	CR
Etmopterus pycnolepis	Etmopteridae	MDB	Chile	LC
Galeus mincaronei	Pentanchidae	MN	Brazil	VU
Gurgesiella dorsalifera	Gurgesiellidae	MDB	Brazil	VU
Hypanus berthalutzae	Dasyatidae	MN	Brazil*	VU
Hypanus marianae	Dasyatidae	MN	Brazil	EN

Scientific name	Family	Habitat	Geography rangy	IUCN
Isogomphodon oxirhynchus	Carcharhinudae	MN	Brazil to Venezuela	CR
Malacoraja obscura	Rajidae	MDB	Brazil	LC
Parmaturus angelae	Pentanchidae	MDB	Brazil	VU
Pseudobatos horkelii	Rhinobatidae	MN	Brazil to Argentina	CR
Rioraja agassizi	Arhynchobatidae	MN, MDB	Brazil to Argentina	VU
Schroederichthys saurisqualus	Scyliorhinidae	MN, MDB	Brazil	VU
Scyliorhinus cabofriensis	Scyliorhinidae	MDB	Brazil	LC
Scyliorhinus ugoi	Scyliorhinidae	MDB	Brazil	LC
Squalus albicaudus	Squalidae	MDB	Brazil	DD
Squalus bahiensis	Squalidae	MDB	Brazil	DD
Squatina varii	Squantinidae	MDB	Brazil	LC
Sympterygia lima	Arhynchobatidae	MN	Chile	LC
Sympterygia acuta	Arhynchobatidae	MN	Brazil to Argentina	CR
Sympterygia bonapartii	Arhynchobatidae	MN, MDB	Brazil to Chile	NT
Triakis acutipinna	Triakidae	MN	Ecuador (mainland)	EN
Urobatis tumbesensis	Urotrygonidae	MN, MDB	Colombia to Peru	VU
Zapteryx brevirostris	Trygonorrhinidae	MN	Brazil to Argentina	EN
Diplobatis colombiensis	Narcinidae	MN	Colombia	VU

Types of habitats: Marine Neritic (MN) and Marine Deep Benthic (MDB); asterisks (*) represent species that occur in island regions; IUCN status: Data Deficient (DD) Least Concern (LC), Vulnerable (VU), Endangered (EN) and Critically Endangered (CR).

Table 1.
Endemic elasmobranch species from South America, accessed through the International Union for Conservation of Nature (IUCN)'s red list of threatened species.

The Western Atlantic Ocean has the highest number of records of endemic species, with the majority attributed to Brazil, which hosts 22 species, and only one representative exclusive each for Argentina and the Falkland Islands. In the portion of the South American continent bordered by the Pacific Ocean, Chile is the country with the highest number of endemic records, cataloging nine species, while Colombia, Ecuador, and the Galapagos Islands together have six representatives. In the assessment of the conservation status of species in South America conducted by the IUCN, nearly half of the endemic marine elasmobranch species (40%) are classified within some category of extinction threat. The majority of the species have been assessed as Least Concern (31.5%), followed by the Vulnerable and Data Deficient categories, representing ~27 and ~16% of the species, respectively. Four species (10.5%) are classified as Endangered, while the Menni's Skate (*Dipturus mennii*) is the only species categorized as Critically Endangered, with its geographic distribution in Brazil.

The type of habitat in which the species occur was also verified, with the same species being present in two types of habitats: Marine Neritic (MN), characterized as a region of the marine environment that is located on the continental shelf, and Marine Deep Benthic (MDB), an environment associated with the ocean floor which may extend beyond the continental shelf. Among the two types of habitats mentioned, most species (n = 16) are restricted to the Marine Deep Benthic habitat. Of the total endemic species evaluated, were verified the occurrence of six species with representatives in island environments (e.g., Lutz's Stingray *Hypanus berthalutzae*, which occurs in the Fernando de Noronha archipelago and in the Rocas Atoll). In general terms, endemism can be caused by physical, climatic, or biological barriers that limit the distribution of a species or separate it from the original group. The more specific the environment, the greater the degree of endemism, that is, the greater the number of endemic species. In this sense, it is of great importance to identify and characterize the areas of endemism in each environment, since these can be included in the selection of priority areas for nature preservation.

2.2 Anthropogenic threats

A series of human activities are currently threatening marine ecosystems and, consequently, resident or migratory species in all oceans. For elasmobranchs, all this pressure, direct or indirect, can strongly affect the populations of rays and sharks. In this sense, aspects related to fishing pressure, pollution and environmental degradation will be detailed and discussed here.

2.3 Fisheries

Shark and ray populations have declined significantly in recent years as a result of extensive human exploitation throughout the world's oceans, putting some species at risk of extinction. These species are captured in some fisheries as target species and/or as bycatch, both by artisanal fishing and by large-scale industrial fishing [9]. Due to the great diversity of habitats of these animals, they end up interacting and are susceptible to a wide variety of fishing gear (e.g., bottom trawling, surface trawling, bottom longline, pelagic longline, and trolling; **Figure 1**).

Figure 1.
Demonstration of the main fishing gears that capture sharks and rays, either as target species or bycatch. (A) trawl net; (B) longline.

In all South American countries, sharks and rays are consumed both locally and for export. For example, Brazil, where sharks and rays are sold as "cação" without identifying the species, is one of the countries that consumes the most sharks. It is also one of the countries that exports the most fins to the Asian market [10].

Although fishing for some species is prohibited, they end up being fished as bycatch and, in many cases, returned to the sea. Bottom trawling is responsible for capturing 19 species of endemic rays, followed by gillnets responsible for catching nine endemic species, while longlines are responsible for catching five species (**Table 2**). Some species are subject to fishing pressure from different fishing gear, such as *Hypanus marianni* and *Pseudobatos horklii* [11]. Endemic rays such as those of the genus *Sympterygia* that are listed as critically endangered by the IUCN and that live close to the substrate, are species captured by artisanal and industrial fishermen who practice bottom trawling throughout their distribution and also interact with other fishing gear [11, 12]. Uncontrolled fishing of some endemic species with low distribution can lead to the decline of these populations and extinction. An example is the ray *Pseudobatos horklii*, which has suffered great fishing pressure throughout its distribution and is classified as critically endangered by the IUCN [11].

Species rays	Trawl	Gillnet	Longline	Recreational	Beach seine
Atlantoraja castelnaui	■				
Dipturus lamillai	■				
Benthobatis kreffti	■				
Dipturus chilensis	■		■		
Dipturus ecuadoriensis					
Dipturus mennii	■	■			
Discopyge castelloi	■	■			
Gurgesiella dorsalifera	■				
Hypanus berthalutzae	■	■	■		
Hypanus marianae	■	■	■		
Malacoraja obscura					
Psammobatis scobina	■		■		
Pseudobatos horkelii	■	■	■	■	■
Rajella eisenhardti					
Rajella sadowskii	■	■			
Rioraja agassizi	■	■			
Sympterygia acuata	■	■	■	■	
Sympterygia bonapartii	■	■	■	■	
Sympterygia lima	■	■			
Urobatis marmoratus	■	■			
Urobatis tumbesensis	■	■			
Zapteryx brevirostris	■	■			

Table 2.
Endemic ray species and the main fishing gear that interacts with them, represented by gray box.

2.4 Pollution

Pollution can be defined as the introduction of any substance or energy into the environment at levels that cause deleterious effects on human health and other living organisms and interfere with the functioning of part or all of the ecosystem [13, 14]. Thus, the accumulation of pollutants in natural environments, especially in aquatic ecosystems, has become one of humanity's greatest concerns in recent decades, since water is an important means of chemical transport, constantly influenced by human activities that occur in terrestrial and river systems [15], in addition to being an excellent solvent.

Pollutants can also have different properties, such as stock pollutants, which include plastics and other non-biodegradable materials, synthetic chemicals, and/or heavy metals, which tend to accumulate in the environment and in living organisms over time and along the food chain [16]. Among the stock pollutants, plastic is a global concern. Most plastic waste reaches aquatic environments through irregular disposal, and estimates indicate that the discharge of plastic waste into rivers, lakes, and oceans reaches 23 million tons per year [17]. Plastic can be carried by rivers, ocean currents, or even pushed by the wind with ease due to its weight and structure, and can accumulate in true floating islands, such as the Pacific Plastic Vortex, which covers approximately 1.5 million km^2 [18]. This is extremely worrying since plastic is a persistent material in the environment and has natural removal rates on the scale of decades to centuries [17]. There are currently several records in the scientific literature on the negative interaction of elasmobranchs with plastic waste in the South Atlantic, especially entanglement, such as that recorded in tiger sharks, *Galeocerdo cuvier* (Péron and Lesueur, 1822) [19], blue sharks, *Prionace glauca* (Linnaeus, 1758) [20], sharpnose sharks, *Rhizoprionodon lalandii* (Valenciennes, 1839) [21] and manta rays, *Mobula birostris* (Walbaum, 1792) and *M. cf. birostris* [22]. Many of these elasmobranch entanglements are caused by ghost fishing gear (74% of animals), followed by polypropylene straps (11% of animals) and other entanglement materials such as circular plastic debris, polyethylene bags and rubber tires (1% of total entangled animals) (**Figure 2**) [23].

Plastic also ends up being mistakenly consumed by elasmobranchs, especially microplastics. In addition to the potential for bioaccumulation and biomagnification, these plastic wastes also interact with chemical pollutants, which adhere to the plastic particles that facilitate the dispersion of these pollutants [24]. The presence of chemical pollutants has also been constantly observed in aquatic environments, which commonly present a mixture of metals, persistent organic pollutants (POPs), pharmaceuticals, cosmetics, personal hygiene products, surfactants and pesticides [24, 25]. These chemical pollutants also have the capacity for bioaccumulation and biomagnification in elasmobranchs, and many can function as endocrine disruptors or be proven to be carcinogenic and mutagenic. Malformations, for example, have been observed in the Spotback Skate, *Atlantoraja castelnaui* (Miranda Ribeiro, 1907) [26] and in blacktip sharks, *Carcharhinus falciformis* (Bibron, 1839) [25] and associated with the presence of chemical pollutants in regions of the South Atlantic. These negative effects resulting from pollution are worrying in the short and long term, as they can interfere with the ability of these animals to compete for food, space and reproductive partners, reducing the persistence capacity of elasmobranch populations, since sharks and rays are strategist K species, which commonly take a long time to reach reproductive age, compared to teleosts and other taxa, and effectively contribute to the maintenance of the size of their populations.

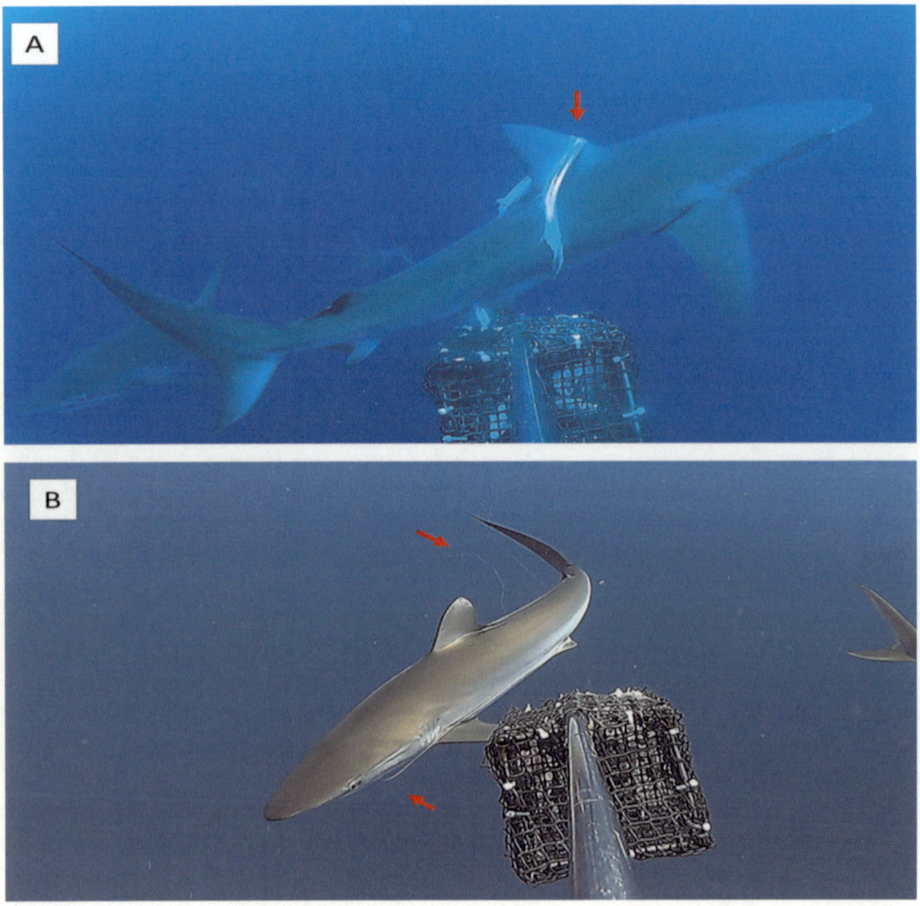

Figure 2.
*(A) Galapagos shark (*Carcharhinus galapagensis, *Snodgrass and Heller, 1905) with a strip of tissue around its body and (B) Silky shark (*Carcharhinus falciformis, *Müller and Henle, 1839) with a hook and line attached to its body recorded through remote underwater video in the southwest Atlantic. Image credits: Camila Araújo, ECOTUBA Project, Federal Rural University of Pernambuco (2024).*

2.5 Environmental degradation

The introduction of pollutants into aquatic environments influences environmental conditions such as dissolved oxygen concentration, pH and turbidity, often reducing the quality of these environments. The discharge of domestic sewage, for example, can lead to an increase in the concentration of nutrients that cause excessive growth of algae, which in turn reduce the passage of light and oxygenation of aquatic environments and often produce toxic compounds such as hydrogen sulfide (H2S) and ammonia (NH3), creating dead zones [15]. In freshwater environments, there are numerous reports of eutrophic environments due to excess nutrients from waste from human activities, and in recent years, the number of reports of macroalgae blooms in marine environments has been increasing, with species of the genus Ulva and the genus Sargassum being the algae commonly involved in the green tides and golden tides, respectively [15]. Another effect of the deposition of nutrients from human activities in aquatic ecosystems is nutrient-induced bioerosion, which, in addition to causing degradation of the coral reef structure, also intensifies the effects of ocean acidification [15] and can sometimes cause the death of coral reefs, areas intensively used by sharks and rays for resting, reproduction and feeding.

Changes in the physical structure of the environment, caused by urbanization and the construction of dams, canals, and ports, and by exploitative activities such as predatory fishing with explosives or trawl nets, as well as gas and oil exploration, modify natural landscapes, often reducing the quantity and quality of available aquatic habitats. Habitat characteristics play a fundamental role in the structuring of aquatic communities, influencing the distribution of species and the presence of mesopredators and top predators such as sharks and rays [27]. Species with highly specialized life histories and limited spatial or environmental distribution are more vulnerable to habitat changes. Many shark and ray species use shallow coastal areas as nurseries to protect their neonates and juveniles from large predators and adverse environmental conditions. Thus, the reduction in the quantity and quality of habitats often compromises feeding, reproduction, and shelter areas, which compromises the recruitment of some species and further contributes to the reduction of biodiversity.

Competition for space and food with invasive alien species also compromises the availability and quality of habitats. Biological invasions have complex direct and indirect impacts, often observed in the long term, when invaders are well-established and have large spatial ranges. Invasive alien species affect the richness and abundance of native species, change the behavior of these animals, alter the phylogenetic diversity among communities, and modify trophic networks, increasing the risk of extinction of native species [28]. Overall, there is little direct evidence that invasive alien species threaten elasmobranchs; however, there are several evidence of how invasive alien species interfere with the trophic dynamics of communities, compromising the food supply and the integrity of the natural ecosystems that elasmobranchs depend on [28].

3. Results

3.1 Alternatives for conservation

As previously mentioned, elasmobranch populations are being directly affected by several anthropogenic threats, as well as marine ecosystems, making the group highly vulnerable to various human activities. In this sense, the development of research aimed at investigating the biology, ecology, movement pattern, and habitat use of sharks and rays is being heavily employed through the use of several non-lethal techniques, in the hope of understanding the population dynamics of sharks and rays and, consequently, contributing to the conservation of the group. Among the various non-lethal methods employed in fisheries science, we highlight the study of blood cells and their relationship with pollutant concentrations in the environment, underwater observation (visual census), investigation of movement patterns and habitat use, through the extensive use of electronic tags (telemetry), and even the observation of animal diversity and behavior through Baited Remote Underwater Video (BRUV) and Drones.

3.2 Ecotoxicology

The vulnerability of elasmobranchs has given rise to a new field of study regarding the concentrations and effects of chemical substances of anthropogenic origin on these species. The bioavailability of these substances, that is, their availability for absorption by organisms, is a risk, especially for aquatic organisms, since these animals live submerged and in permanent contact with a mixture of chemical substances [24]. These bioavailable substances can be absorbed through gill respiration and the ingestion of food and water, in the case of marine fish, or through the tegument, the tissue that covers the body of these animals, bioaccumulating in these organisms over time, or biomagnifying through the food chain, especially in top predators, such as sharks and aquatic mammals [29].

The bioaccumulation and biomagnification of substances in elasmobranchs occurs in different tissues, especially in adipose tissues and organs such as the liver, the main organ involved in the detoxification process [30]. Sharks, in particular, have large, fat-rich livers and are very susceptible to the accumulation of lipophilic substances, such as metals and persistent organic pollutants (POPs), and tend to have higher metal concentrations compared to teleosts of similar trophic levels [24]. In South America, high concentrations of POPs were observed in the liver and muscle of blue sharks, *Prionace glauca*, at concentrations higher than those permitted for human consumption by the European Commission (maximum of 200 ng g−1 ww), as well as high concentrations of Mn, Cd, Fe, and Cu [31]. Maternal transfer of polycyclic aromatic hydrocarbons (PAHs) has also been observed in elasmobranchs in the South Atlantic, more precisely in the endemic Brazilian guitarfish *Pseudobatos horkelii*, which presented an average rate of PAHs discharged to offspring of 13%, with low molecular weight PAHs presenting the highest rates of maternal transfer [32].

Responses of organisms to contact with these exogenous substances have been used as biomarkers of the health of these animals and the environment in which they live. In South America, the activities of some of these biomarkers have begun to be investigated in sharks and rays, such as studies on the activity of cholinesterases (ChE), enzymes that play an essential role in neuronal and motor functions, in brain and muscle tissues of blue sharks. The brain of blue sharks contains atypical ChE, exhibiting mixed properties of acetylcholinesterase (AChE) and butyrylcholinesterase (BChE), differently from bony fish, which implies different detoxification mechanisms [31]. The activity of the enzyme ethoxyresorufin-O-deethylase (EROD), an essential biotransformation enzyme for the detoxification of a wide range of xenobiotics, was also investigated in the catshark, *Schroederichthys chilensis*, captured in the South Pacific Ocean, in conjunction with the determination of fluorescent aromatic compounds, to indicate the occurrence of exposure to PAHs. EROD activity and fluorescent aromatic compounds showed significant differences between study areas, with catsharks captured in PAH-contaminated areas presenting EROD activity values 4.2 and 2.4 times higher than the uncontaminated reference study area [33]. The induction of EROD can produce deleterious side effects, such as DNA-PAH adducts (binding between DNA molecules and PAHs), which are crucial factors in the development of cancer in fish and are also linked to deficiencies in both reproduction and the immune system.

Genotoxic responses have also been observed through blood samples collected by non-lethal methods in tiger sharks, lemon sharks, and nurse sharks, sampled in a conservation area in the South Atlantic. Significant variations among species, which are also part of different trophic levels, were observed and associated with the presence of pollutants in the water column, suggesting that contact with pollutants such as metals and surfactants, even in small quantities, can cause genotoxic effects in sharks (**Figure 3**) [34]. These biomarkers do not provide information about impacts at more complex organizational levels, such as populations, communities and/or ecosystems, but they enable the detection of harmful effects early on at basal levels, which occur

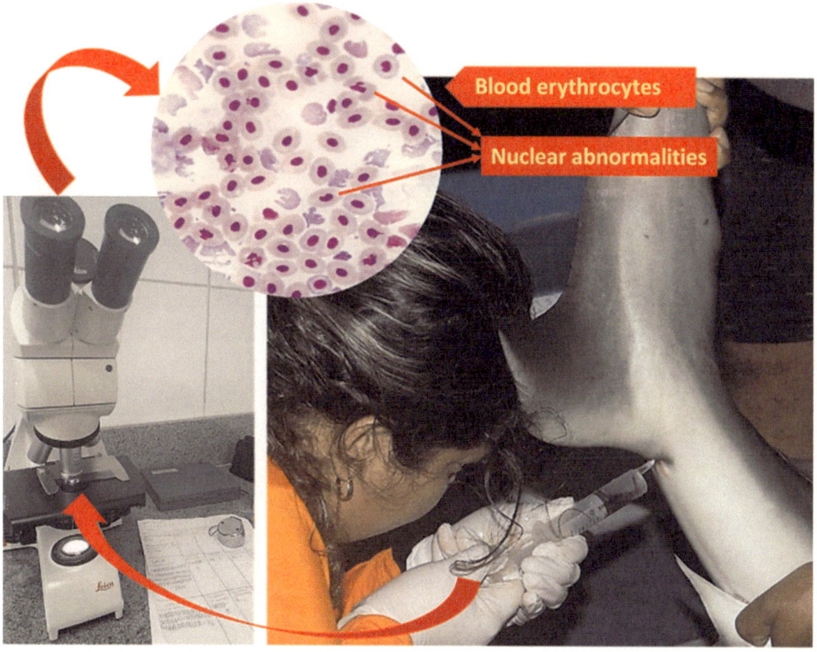

Figure 3.
The analysis of genomic damage has contributed to the understanding of the genotoxic responses of sharks in the South Atlantic. Image credits: Camila Araújo, ECOTUBA Project, Federal Rural University of Pernambuco (2024).

on a scale of minutes to weeks, and it is essential to have a correlation between their frequencies and concentrations with the intensity of exposure to the stressor [35].

Relationships between metal and metalloid concentrations and the frequency of genomic damage have also been investigated in tiger sharks, nurse sharks, and lemon sharks, with Al, As, and Zn showing a significant effect on the frequencies of genomic damage for all three species. Interspecific variations were also observed in this study, with Zn influencing the frequencies of binucleate cells and Al the frequencies of total damage and micronuclei in tiger sharks and lemon sharks, and As influencing the frequencies of binucleate cells and notched nucleus in nurse sharks, while showing a strong and positive correlation with most of the metals analyzed [36]. The high frequency of genomic damage over a long period of time is worrying because it can lead to disturbances in development and reproduction that can compromise the fitness of the organisms.

In general, there are still few studies focusing on pollutant levels and their effects on elasmobranchs, with most studies focusing on economically important sharks and rays, highlighting the lack of studies on species with low economic value and critical conservation status. There are also few studies using non-invasive methodologies, with muscle and liver being the most analyzed organs in current ecotoxicological studies. Therefore, the knowledge about the impacts of pollution on the health of these animals is urgent, since pollution of aquatic environments is a growing problem, especially in developing countries, constituting an additional stressor for many elasmobranch species that are already under threat of extinction due to fishing exploitation and habitat loss [37]. Thus, understanding the negative effects associated with pollution can greatly contribute to the implementation of public policies aimed at the conservation of elasmobranchs and their essential habitats.

3.3 Visual census

Visual census is a method used to observe individuals in an aquatic environment directly, through free diving or scuba diving [38]. Visual censuses serve to collect relevant information on spatial ecology, abundance, density, distribution, and population structure, distribution by sex and size, in addition to providing information on behavior in the wild [38, 39]. Visual census has been widely used, especially for endangered species or in marine protected areas, as they do not involve capture [38, 39]. Visual census can be performed in transects, intensive search, or stationary points depending on the objective and species studied. In transects, divers follow a parallel line, noting all the individuals observed. Transects can be used in different environments such as shipwrecks and reef environments [38].

During the visual census, it is possible to evaluate the different behaviors performed by sharks and rays (e.g., resting, swimming, feeding, and reproduction) and correlate them with the environment in which they are being monitored (e.g., type of substrate, depth, temperature, and tide, among others). The Atoll Biological Reserve (Brazil) is an ecological sanctuary for several migratory and resident species. Since it is a marine protected area, where only research activities are carried out, a series of studies involving non-lethal methods have already been carried out to investigate the populations of sharks and rays (e.g., Lemon Shark, *Negaprion brevirostris*, *G. cirratum*, and *H. berthalutzae*). The Rocas Atoll is an island ecosystem in the Western Atlantic Ocean, which is strongly linked to the sea regimes. During periods when the tide is rising or falling, there is intense movement of sharks and rays, either toward the interior or exterior of the Rocas Atoll.

H. berthalutzae is a species of stingray endemic to South America (Brazil), which has been extensively studied using non-lethal methods in the Rocas Atoll Biological Reserve and the Fernando de Noronha Archipelago [38]. To expand the knowledge available in the literature on Lutz's stingray regarding population structure, behavior, and seasonality, a monitoring period of approximately one decade was carried out at Rocas Atoll. The visual census was performed only inside the atoll, where it was verified that most of the time, the specimens remained at rest, camouflaged in the sandy substrate. Individuals moving on rocky substrates were observed less frequently (**Figure 4**). A predominance of females over males was also observed (sex ratio of 1 female: 0.14 males), in addition to adult individuals predominating in the underwater observations. The results reported for the Rocas Atoll were different from the data obtained in the Fernando de Noronha Archipelago. In this location, there is a greater predominance of young individuals in shallow areas, in addition to a greater balance in the proportion between males and females.

In general terms, the visual census has proven itself over the years to be an extremely efficient, low-cost, and highly recommended technique for investigating shark and ray populations on oceanic islands, clear water, and shallow areas, favoring the implementation of the methodology and providing data acquisition on wild animals, thus contributing to the conservation of populations.

3.4 BRUVS and drones

Recent technological advancements have significantly enhanced the development of non-lethal and non-invasive sampling methods for assessing fish assemblages. This is particularly beneficial in hard-to-reach environments, areas that present risks to samplers, and ecological sanctuaries with high preservation values. Furthermore,

Figure 4.
Underwater observations (visual census) of Lutz's stingray H. berthalutzae, *carried out in the Biological Reserve of Atol das Rocas and the Archipelago of Fernando de Noronha (AFN). (A) individual resting on the sandy substrate, in Atol das Rocas (image credits: Ilka Branco); (B) active specimen, moving on the rocky substrate, in Atol das Rocas (image credits: Mariana gabriela); (C) active individual on the sandy substrate, with algae, in AFN (image credits: Sibele Mendonça); (D) female specimen in Atol das Rocas (image credits: Ilka Branco).*

research that minimizes impacts on organisms and their habitats is particularly encouraged for endangered species and areas identified as hotspots, where the removal of animals from their environment could have significant effects on their populations, especially in the case of endemic species. In this context, the use of underwater video cameras for inventorying and monitoring fauna and their habitats has been on the rise, as they provide permanent records that do not impact the natural environment [40].

Baited Remote Underwater Video Surveys (BRUVs) have yielded highly satisfactory results on fish and elasmobranch research assemblages in various marine ecosystems [41]. This equipment can be constructed using readily available materials, featuring a relatively simple design that consists of a compact video camera housed in a waterproof case, attached to a base made in various shapes, and directed to a box with baits.

The number of cameras associated with the structure will depend on the objectives of the investigations: Single-BRUVs are equipped with a single camera and are less effective at estimating organism sizes compared to Stereo-BRUVs, which are constructed with two cameras. The latter not only provides more accurate length estimates but also expands the field of view of the monitored area [42]. The cameras capture underwater images of the species surrounding the equipment, primarily attracted by the bait contained in the boxes, although some animals may also approach the structure out of curiosity about the new presence.

Similar to underwater visual censuses, BRUVs enable the assessment of biodiversity, abundance, and spatial distribution of species, as well as the inference of important data regarding the population structure of a region [43], all without the presence of samplers in the water. Consequently, recent studies have focused on species behavior in their natural habitats using the autonomous recordings obtained from BRUVs, thereby generating much more reliable data on the natural behaviors of these species [44].

The BRUVs have been successfully utilized, particularly in Marine Protected Areas (MPAs), especially in fully protected zones, with the aim of monitoring biodiversity and assessing the effectiveness of these locations [45]. On the other hand, in areas with a high degree of environmental impact, BRUVs are employed to measure the extent of habitat degradation reflected in the species community at each site. In anthropogenically impacted areas of Pernambuco, Brazil, Lutz's stingray (*Hypanus berthalutzae*), an endemic species vulnerable to extinction, was documented in images produced by BRUVs in the region [46]. The distribution of Lutz's stingray extends to the oceanic islands of the Atol das Rocas and the Fernando de Noronha archipelago, which are important marine protected areas (MPAs) for the conservation of this species (**Figure 5**).

Furthermore, the use of BRUVs allows for a greater number of samples to be collected in a shorter time frame and facilitates research in deeper environments due to the autonomy of the equipment, eliminating the need for divers to conduct the work. Globally, the use of BRUVs in scientific studies is on the rise due to their operational advantages and low cost.

Figure 5.
*Lutz's stingray (*Hypanus berthalutzae*): A Brazilian endemic species recorded using BRUVs in the state of Pernambuco, Brazil.*

However, it is important to note that, like all research methods capable of estimating species abundance and diversity, their use should be coupled with different methodologies to mitigate the sampling biases inherent to each method, as well as to allow for comparisons regarding the precision of the results obtained [47].

Within the same framework of faunal inventory and monitoring of underwater regions using autonomous equipment, remotely operated vehicles (ROVs) represent a more technically advanced data collection tool. In addition to capturing images with an attached video camera, ROVs are equipped with sensors and tools to perform specific tasks, such as collecting samples. The ability of ROVs to operate at depths inaccessible to divers, combined with their capacity to enter confined and narrow spaces, makes this equipment extremely valuable across various sectors, including scientific research.

In marine fauna investigations, ROVs differ from BRUVs in that they do not rely on bait as an attractant. Instead, they are capable of performing movements and actions controlled by an operator remotely, which is particularly beneficial for investigating species in deep regions where endemic species may be identified that would likely go unseen by other methods [48]. However, due to their greater autonomy, efficiency in data generation, and application across various research fronts, the cost of acquiring and operating ROVs remains a barrier to the development of academic research.

Drones, or unmanned aerial vehicles (UAVs), have gained popularity due to their ability to efficiently capture aerial imagery and the wide range of price points available for acquisition. Consequently, these technologies offer a broad array of functionalities that can be applied in various activities and environments.

Drones are model aircraft equipped with an autopilot system and can be either remotely controlled by an operator or fly autonomously, following predefined routes or control programs. Although the images generated by drones provide a perspective from above the water, these devices have become an important tool in research for the identification and monitoring of marine megafauna species. Additionally, they capture behaviors in natural habitats, area usage patterns, and estimates of population structure within ecosystems.

Large pelagic elasmobranchs, such as mobulids and whale sharks, are easily detected by these devices when they migrate to the ocean's surface [49]. Drones can access areas that would be extremely challenging to reach, particularly in regions with difficult navigation or usage restrictions, and they also have lower logistical costs compared to some traditional research methods. In addition, drones are an essential tool for species conservation due to their ability to monitor the recovery of degraded areas and to document illegal activities in marine protected areas. This capability aids in identifying issues and facilitating the prosecution of those responsible. Technological innovations have not only reduced the size of these devices but also lowered acquisition costs, allowing for an expansion in the number of studies conducted with drones worldwide [49].

3.5 Electronic tagging

The movement patterns of aquatic organisms provide a suitable understanding of ecological and evolutionary processes that occur across different spatial and temporal scales. Sharks and rays should optimize their movements to balance the energetic and life history requirements with the costs of movement, such as reproductive and trophic migrations [50]. The choice of specific habitats within an ecosystem can be crucial to the survival and reproductive success of a species. In the early 1950s, the use of conventional (plastic) tags made it possible to investigate the movement of elasmobranchs. However, due to the limitations of the data obtained by tag-recapture work, biotelemetry techniques were developed in the hope of expanding monitoring both on a geographic scale and in greater detail of habitat use.

Acoustic telemetry is a frequently used method to assess the presence and movement information from aquatic animals, including skates and rays. This tool involves the use of transmitters implanted in the animals to be tracked and a hydrophone (usually called to as an acoustic receiver) that will have the capacity to detect the signal emitted by the transmitters. The data from acoustic telemetry are usually collected in two different ways, be it for active tracking or passive tracking, and each technique has its singular advantages and limitations [51].

Active tracking normally uses specific transmitters that emit continuous acoustic pulses and the signal is monitored by a receiver and hydrophone carried onboard the tracking vessel. Tracking invariably takes place over short durations of a few hours to a few days. The main objective of this technique is to provide detailed, fine-scale (m) tracks of animals by following the signal from the transmitter. The severe nature of

Figure 6.
Tagging techniques in sharks and rays. (A) External tagging with an acoustic transmitter on the tail of the H. berthalutzae *(image credits: Diogo Nunes); (B) tagging with a satellite transmitter on the pectoral fin, in Atol das Rocas (image credits: Ilka Branco); (C) and (D) internal tagging (surgery) in the Tiger shark* Galeocerdo cuvier, *in the Fernando de Noronha Archipelago (image credits: Camila Araújo, ECOTUBA Project, Federal Rural University of Pernambuco (2024).*

this fieldwork means that it can normally only be conducted for a brief time, being therefore, essential that a rapid method of transmitter attachment is used to minimize stress and, consequently, not promote unusual behaviors by the study animal during the tracking [52]. Passive telemetry is based on the use of an array of acoustic receivers (omni-directional hydrophone) installed in underwater stations capable of detecting the presence of an individual, with a coded transmitter, for a long period of time. However, it is not possible to know the location of the individual, which carries the acoustic transmitter, if it is not close to the array of receivers.

Transmitters used in acoustic telemetry (active or passive) can be attached to sharks and rays externally or internally. External tagging can be done in several ways. For sharks, it is extremely usual to attach transmitters to the dorsal fin of the specimen, while for rays, this attachment can be done to the pectoral fin, tail, or even the blowhole (**Figure 6a**). Internal tagging involves a surgical procedure in which a small incision (~4 cm) is made in the animal's abdominal cavity, the transmitter is inserted, and the incision is sutured (**Figure 6c** and **d**). The choice of tagging technique (internal or external) should be based on the ecology of the species. In this sense, it is very important to evaluate the biology and behavior of the animal to choose the best form of tagging, ensure the retention of the transmitter, and consequently, the success of the monitoring.

An additional tool, which is widely used in research around the world, is satellite telemetry (**Figure 6b**). Individual movement patterns are recorded remotely while the tag is fixed to a specimen, and the data are transmitted *via* satellite. The satellite transmitters can record data of estimated geographic position, depth, and ambient temperature [53]. The main advantage of this technique is the possibility of monitoring an individual, for a long time, on a large spatial scale, without the need for the installation of receivers. The tagging and monitoring techniques reported here have revolutionized studies of behavior and habitat use for elasmobranchs, enabling the acquisition of valuable data using a non-lethal method.

4. Discussion

4.1 Challenges and perspectives

The endemic species of elasmobranchs (sharks and rays) in South America face numerous challenges and opportunities for conservation. Here's an overview of the key challenges and perspectives. Many species are targeted for their meat, fins, and cartilage, leading to significant population declines, and unsustainable fishing practices exacerbate this problem. Combined with this, coastal development, pollution, and habitat degradation (such as the deforestation of mangroves and the destruction of coral reefs) threaten critical breeding and nursery habitats for these species. Rising sea temperatures and ocean acidification can impact the health of marine ecosystems by affecting food availability and habitat suitability for elasmobranchs, which, without enforcement of fishing regulations and illegal fishing activities, contribute to the decline of endemic species. Another key factor is that many endemic species lack comprehensive research data, hampering effective conservation strategies, and this knowledge gap can make it difficult to assess population health and trends. Finally, as coastal communities expand, interactions between humans and elasmobranchs can lead to conflicts, further complicating conservation efforts.

From perspectives, we can initially mention the conservation initiatives that culminate in the growing awareness and defense of marine conservation aimed at protecting elasmobranch habitats and populations, in addition to the creation of marine protected areas (MPAs) as crucial for the preservation of biodiversity. Promoting sustainable fishing practices can also help restore populations of overexploited species, and certification programs and community-led management initiatives are gaining momentum. Investment in research can help fill knowledge gaps and the development of citizen science projects, and partnerships between governments, NGOs, and academic institutions can improve monitoring efforts. In this sense, local community involvement in conservation efforts promotes management and can lead to more effective management of marine resources, as well as strengthening legal structures and international agreements can help protect elasmobranchs, combined with collaborative efforts between countries. South American species are essential for transboundary species management.

Addressing the challenges faced by endemic elasmobranchs in South America requires a multifaceted approach that includes sustainable practices, strong legal frameworks, community engagement, and dedicated research efforts. By prioritizing these areas, it is possible to create a more favorable environment for the survival of these unique and vital species.

5. Conclusions

Almost half of the endemic sharks and rays in South America are classified within some category of extinction threat. Therefore, non-lethal research methods are strongly encouraged for studies involving this group. Additionally, demystifying the consumption of elasmobranchs in South America, particularly those lacking species identification labels, is also aimed at curbing the illegal trade of threatened species. These measures offer hope for the conservation of shark and ray populations, which are essential for maintaining the balance of marine ecosystems.

Acknowledgements

This chapter is dedicated to Professor Dr. Fábio Hazin for all his contributions to Fishery Science and for transforming the lives of all his students with the best opportunities. The authors are thankful to the Departamento de Pesca e Aquicultura at the Universidade Federal Rural de Pernambuco for all the professional support.

Conflict of interest

The authors declare no conflict of interest.

Author details

Ilka Branco-Nunes[1]*, Camila Araújo[1], Emmanuely Ferreira[1], Sibele Mendonça[1], Natalia Alves[1,2], Diogo Nunes[1], Danielle Viana[1] and Paulo Oliveira[1]

1 Laboratory of Fish Ethology, Departamento de Pesca e Aquicultura, Universidade Federal Rural de Pernambuco, Rua Dom Manuel de Medeiros, Dois Irmãos, Pernambuco, Brazil

2 Laboratory of Marine Bioecology, Universidade Federal do Espírito Santo, Vitória, Espírito Santo, Brazil

*Address all correspondence to: iilkabranco@hotmail.com

References

[1] Nelson JS, Grande TC, Wilson MV. Fishes of the World. 5th ed. Hoboken: Wiley; 2016. 711p. DOI: 10.1002/9781119174844

[2] Whitenack L, Kim S, Sibert EC. Bridging the gap between chondrichthyan paleobiology and biology. In: Carrier JC, Simpfendorfer CA, Heithaus MR, Yopak KE, editors. Biology of Sharks and their Relatives. 2nd ed. Boca Raton: CRC Press; 2022. pp. 1-29. DOI: 10.1201/9781003262190

[3] Paig-Tran EWM, Porter ME, Ferry LA, Whitenack LB. How to build a shark biomechanics and bioinspiration. In: Carrier JC, Simpfendorfer CA, Heithaus MR, Yopak KE, editors. Biology of Sharks and their Relatives. 2nd ed. Boca Raton: CRC Press; 2022. pp. 59-103. DOI: 10.1201/9781003262190

[4] White WT, O'Neill HL, Naylor GJP. Taxonomy and diversity of extant elasmobranchs. In: Carrier JC, Simpfendorfer CA, Heithaus MR, Yopak KE, editors. Biology of Sharks and their Relatives. 2nd ed. Boca Raton: CRC Press; 2022. pp. 31-57. DOI: 10.1201/9781003262190

[5] Harry AV, Smart JJ, Pardo SA. Understanding the age and growth of chondrichthyan fishes. In: Carrier JC, Simpfendorfer CA, Heithaus MR, Yopak KE, editors. Biology of Sharks and their Relatives. 2nd ed. Boca Raton: CRC Press; 2022. pp. 178-202. DOI: 10.1201/9781003262190

[6] Cortés E. Life history patterns and correlations in sharks. Reviews in Fisheries Science. 2000;**8**:299-344. DOI: 10.1080/10408340308951115

[7] Merwe AB, Lyons K, Kacev D, Feldheim K. Elasmobranch mating systems. In: Carrier JC, Simpfendorfer CA, Heithaus MR, Yopak KE, editors. Biology of Sharks and their Relatives. 2nd ed. Boca Raton: CRC Press; 2022. pp. 203-222. DOI: 10.1201/9781003262190

[8] Heupel MR, Knip DM, Simpfendorfer CA, Dulvy NK. Sizing up the ecological role of sharks as predators. Marine Ecology Progress Series. 2014; **495**:291-298. DOI: 10.3354/meps10597

[9] Pacoureau N, Rigby CL, Kyne PM, et al. Half a century of global decline in oceanic sharks and rays. Nature. 2021; **589**:567-571. DOI: 10.1038/s41586-020-03173-9

[10] Barreto RR, Bornatowski H, Motta F, Santander-Neto J, Vianna GMS, Lessa R. Rethinking use and trade of pelagic sharks from Brazil. Marine Policy. 2017; **85**:114-122

[11] Pollom R, Barreto R, Charvet P, Faria V, Herman K, Marcante F, Rincon G.. The IUCN Red List of Threatened Species. 2020. DOI: 10.2305/IUCN.UK.2020-3.RLTS.T45925A10 4130004.en

[12] Dulvy NK, Acuña E, Bustamante C, Concha F, Herman K, Velez-Zuazo X. Sympterygia lima. In: The IUCN Red List of Threatened Species. 2020, 2020; e. T44643A124436532. DOI: 10.2305/IUCN.UK.2020-3.RLTS.T44643A124436532.en

[13] GESAMP Joint IMCO/FAO/UNESCO/WMO Group of Experts on the Scientific Aspects of Marine Pollution (GESAMP). Rep. 1st Sess. London: IMCO; 1969. p. 28

[14] ACIESP. Glossário de Ecologia. São Paulo: Publicação ACIESP. N° 57; 1987; 271p

[15] Reichelt-Brushett A. Marine pollution in context. In: Marine Pollution–Monitoring, Management and Mitigation. Cham: Springer Nature Switzerland; 2023. pp. 1-22

[16] Garai P, Banerjee P, Mondal P, Saha NC. Effect of heavy metals on fishes: Toxicity and bioaccumulation. Journal of Clinical Toxicology. 2021;**S**:18

[17] MacLeod M, Arp HPH, Tekman MB, Jahnke A. The global threat from plastic pollution. Science. 2021;**373**(6550):61-65

[18] Gomes CA. O plástico, a nova praga ambiental. Medio Ambiente and Derecho: Revista electrónica de derecho ambiental. 2019;**35**:1-12

[19] Afonso AS, Fidelis L. The fate of plastic-wearing sharks: Entanglement of an iconic top predator in marine debris. Marine Pollution Bulletin. 2023;**194**: 115326

[20] Barreto R et al. Macro-debris ingestion and entanglement by blue sharks (Prionace glauca Linnaeus, 1758) in the temperate South Atlantic Ocean. Marine Pollution Bulletin. 2019;**145**: 214-218

[21] Sazima I et al. Plastic debris collars on juvenile carcharhinid sharks (*Rhizoprionodon lalandii*) in Southwest Atlantic. Marine Pollution Bulletin. 2002;**44**(10):1149-1151

[22] Bucair N et al. Underestimated threats to manta rays in Brazil: Primacies to support conservation strategies. Global Ecology and Conservation. 2021; **30**:e01753

[23] Parton KJ, Galloway TS, Godley BJ. Global review of shark and ray entanglement in anthropogenic marine debris. Endangered Species Research. 2019;**39**:173-190

[24] Alves LM et al. Elasmobranchs as bioindicators of pollution in the marine environment. Marine Pollution Bulletin. 2022;**176**:113418

[25] Araújo CBBD. Genotoxicidade comparativa de grandes peixes pelágicos em duas unidades de conservação do Atlântico Equatorial: Arquipélago de Fernando de Noronha e Arquipélago de São Pedro e São Paulo [Thesis]. Recife, Brazil: Programa de Pós-Graduação em Biologia Animal, Universidade Federal de Pernambuco; 2024. p. 115

[26] Hauser-Davis RA, Barbosa-Filho ML, Pereira LHSDS, Lopes CA, Moreira SC, Rocha RC, et al. First record of a morphologically abnormal and highly metal-contaminated Spotback skate *Atlantoraja castelnaui* (Rajiformes: Arhynchobatidae) from southeastern Rio de Janeiro, Brazil. Journal of Threatened Taxa. 2020;**12**(11): 16510-16520

[27] Araújo CBBD. Estruturação espacial e relações tróficas das assembleias de peixes do Atol das rocas, o único atol do Atlântico Sul [Dissertation]. Recife, Brazil: Programa de Pós-Graduação em Ecologia, Universidade Federal Rural de Pernambuco; 2017. p. 88

[28] Pyšek P, Hulme PE, Simberloff D, Bacher S, Blackburn TM, Carlton JT, et al. Scientists' warning on invasive alien species. Biological Reviews. 2020; **95**(6):1511-1534

[29] Tiktak GP et al. Are concentrations of pollutants in sharks, rays and skates (*Elasmobranchii*) a cause for concern? A systematic review. Marine Pollution Bulletin. 2020;**160**:111701

[30] Hauser-Davis RA et al. Metal concentrations and metallothionein

metal detoxification in blue sharks, *Prionace glauca L.* from the Western North Atlantic Ocean. Journal of Trace Elements in Medicine and Biology. 2021; **68**:126813

[31] Alves LM, Lemos MF, Correia JP, da Costa NA, Novais SC. The potential of cholinesterases as tools for biomonitoring studies with sharks: Biochemical characterization in brain and muscle tissues of Prionace glauca. Journal of Experimental Marine Biology and Ecology. 2015;**465**:49-55

[32] Martins MF, Costa PG, Bianchini A. Maternal transfer of polycyclic aromatic hydrocarbons in an endangered elasmobranch, the Brazilian guitarfish. Chemosphere. 2021;**263**: 128275

[33] Fuentes-Rios D et al. EROD activity and biliary fluorescence in *Schroederichthys chilensis* (Guichenot 1848): Biomarkers of PAH exposure in coastal environments of the South Pacific Ocean. Chemosphere. 2005; **61**(2):192-199

[34] Araújo C et al. Comparative genomic damage among three shark species with different habits: Sublethal impacts of human origin in a protected island environment in the South Atlantic. Marine Pollution Bulletin. 2023;**191**: 114924

[35] Adam ML et al. Espécies sentinelas: Monitoramento Ambiental com base em biomarcadores de efeito cito-e genotóxico. Revista CEPSUL - Biodiv e Conser Marinha. 2023;**12**:e20230002-e20230002

[36] Araújo CBB, de Mendonça SA, de Lima Viana D, da Fontoura Martins M, Costa PG, Bianchini A, et al. Effects of blood metal (loid) concentrations on genomic damages in sharks. Environmental Pollution. 2024;**359**: 124569

[37] Wosnick N, Niella Y, Hammerschlag N, Chaves AP, Hauser-Davis RA, da Rocha RCC, et al. Negative metal bioaccumulation impacts on systemic shark health and homeostatic balance. Marine Pollution Bulletin. 2021; **168**:112398

[38] Branco I. Ecologia da raia, Dasyatis americana (Hildebrand and Schroeder, 1928), na região metropolitana do Recife - PE e na ReBio Atol das Rocas - Brasil 2015 [thesis]. Pernambuco, Brazil: Federal Rural University of Pernambuco; 2015

[39] Mendonça SA, Macena BCL, Araujo CBB, Bezerra NPA, Hazin FHV. Dancing with the devil: Courtship behaviour, mating evidences and population structure of the *Mobula tarapacana* (myliobatiformes: Mobulidae) in a remote archipelago in the equatorial mid-Atlantic Ocean. Neotropical Ichthyology. 2020;**18**(3): e200008. DOI: 10.1590/1982-0224-2020-0008

[40] Struthers DP, Danylchuk AJ, Wilson ADM, Cooke SJ. Action cameras: Bringing aquatic and fisheries research into view. Fisheries. 2015;**40**(10): 502-512. DOI: 10.1080/03632415.2015.1082472

[41] Whitmarsh SK, Fairweather PG, Huveneers C. What is big BRUVver up to? Methods and uses of baited underwater video. Reviews in Fish Biology and Fisheries. 2017;**27**:53-73. DOI: doi.org/10.1007/s11160-016-9450-1

[42] Schramm KD, Harvey ES, Goetze JS, Travers MJ, Warnock B, Saunders BJ. A comparison of stereo-BRUV, diver

operated and remote stereo-video transects for assessing reef fish assemblages. Journal of Experimental Marine Biology and Ecology. 2020;**524**: 151273

[43] Rolim FA, Rodrigues PF, Langlois T, Neves LM, Gadig OB. A comparison of stereo-videos and visual census methods for assessing subtropical rocky reef fish assemblage. Environmental Biology of Fishes. 2022;**104**:413-429. DOI: 10.1007/s10641-022-01240-w

[44] Parton KJ, Doherty PD, Parrish M, Shearer P, Myrick K, Shipley ON, et al. Opportunistic camera surveys provide insight into discrete foraging behaviours in nurse sharks (*Ginglymostoma cirratum*). Environmental Biology of Fishes. 2023;**106**:19-30. DOI: 10.1007/s10641-022-01366-x

[45] Rolim FA, Langlois T, Rodrigues PFC, et al. Network of small no-take marine reserves reveals greater abundance and body size of fisheries target species. PLoS One. 2019;**14**: e0204970

[46] Bezerra NPA, Correa AL, Hazin FHV, Garla R. First use of baited remote underwater video stations to assess fish diversity in the metropolitan region of Recife, northeastern Brazil. Revista de Biologia Tropical. 2022;**70**(1): 332-347. DOI: 10.15517/rev.biol.trop..v70i1.45915

[47] Rhodes N, Wilms T, Baktoft H, Ramm G, Bertelsen JL, Flávio H, et al. Comparing methodologies in marine habitat monitoring research: An assessment of species-habitat relationships as revealed by baited and unbaited remote underwater video systems. Journal of Experimental Marine Biology and Ecology. 2020;**526**:151315

[48] Vieira RP, Cunha MR. In situ observation of chimaerid speciesin the Gorringe Bank: New distribution records for the north-East Atlantic Ocean. Journal of fish biology. 2014;**85**(3): 927-932

[49] Frixione MG, García MD, Gauger MF. Drone imaging of elasmobranchs: Whale sharks and golden cownose rays co-occurrence in a zooplankton hot-spot in southwestern sea of Cortez. Food Webs. 2020;**24**: e00155

[50] Humphries NE, Sims DW. Optimal foraging strategies: Lévy walks balance searching and patch exploitation under a very broad range of conditions. Journal of Theoretical Biology. 2014;**358**:179-193. DOI: DOI10.1016/j.jtbi.2014.05.032

[51] Mull CG, Andrzejaczek S, Udyawer V, Dwyer RG. Advances in methods, understanding, and applications of elasmobranch movement ecology. In: Carrier JC, Simpfendorfer CA, Heithaus MR, Yopak KE, editors. Biology of Sharks and their Relatives. 2nd ed. Boca Raton: CRC Press; 2022. pp. 358-399. DOI: 10.1201/9781003262190

[52] Speed CW, Field IC, Meekan MG, Bradshaw CJA. Complexities of coastal shark movements and their implications for management. Marine Ecology Progress Series. 2010;**408**:275-293. DOI: 10.3354/meps08581

[53] Branco-Nunes I, Veras D, Oliveira P, Hazin F. Vertical movements of the southern stingray, *Dasyatis americana* (Hildebrand and Schroeder, 1928) in the biological reserve of the Rocas Atoll, Brazil. Latin American Journal of Aquatic Research. 2016;**44**:216-227. DOI: 10.3856/vol44-issue2-fulltext-3

Chapter 6

Endemic Fish Species in the Macedonian Fresh Water

Trajche Talevski, Marina Talevska and Aleksandra Talevska Leshoska

Abstract

Freshwater ecosystems (rivers and lakes) in Macedonia belong to two ecoregions. These are Eco region 420 (Southeastern Adriatic basin, Prespa and Ohrid lakes) and Eco region 422 (Vardar Basin with Dojran Lake). The Macedonian fish fauna consists of about 80 species of fish that live in Macedonian lakes and rivers. In large natural lakes (Prespa Lake, Ohrid Lake, and Dojran Lake), about 60 species of fish belonging to 15 families have been described, while other fish species are distributed in rivers and artificial lakes. Fish populations in Macedonia can be divided into autochthonous, nonnative, and invasive species. Twenty-two autochthonous and 7 introduced fish species were described in Lake Ohrid; 11 autochthonous and 12 introduced species were described in Prespa Lake; and 15 autochthonous and 7 introduced fish species were described in Dojran Lake. Other species that live in high mountain rivers are mostly autochthonous. Of the indigenous fish species, a large number of fish species in Macedonia are endemic and live in large natural lakes and rivers. The IUCN Red List is considered the most objective and authoritative system for classifying species in terms of extinction risk. According to the IUCN Red List, fish from Macedonian waters belong to the following categories: LC—35 species of fish, CR—3 species, EN—2 species, VU—9 species, NT—1 species, NE—1 species, DD—10 species, IN—18 species, and 1 species not categorized on the IUCN Red List.

Keywords: catchment area, water body, endemic fish species, natural and artificial lakes, rivers, IUCN Red List, danger to endemic fish populations

1. Introduction

The freshwater fish fauna of the western and central regions of the Balkan Peninsula, encompassing the Balkan nations, has evolved into its present form and composition following the last Pleistocene glaciation. Notably, during the Ice Age of the Pleistocene, much of the Balkan Peninsula remained free from glacial cover (with the exception of isolated ice masses at higher elevations), creating refugia where various fish species thrived. After the conclusion of the Würm glaciation, these species migrated northward, primarily through the Danube River. The presence of numerous isolated habitats facilitated the emergence of endemic species, particularly in the southern European region (Adriatic Sea basin). Some of these species are classified as paleoendemites, particularly those associated with ancient basins like Lake Ohrid, which dates back to the Pliocene Tertiary. The age of this fauna is corroborated by paleontological evidence from the Northern Balkans and Central Europe, which mainly includes extinct species of recent genera that continue to inhabit these areas today [1].

The indigenous freshwater fish fauna of the central and Western Balkans, which includes water bodies situated between the southern slopes of the Alps in the northwest, the southern edges of the Pannonian Plain in the north, the southern branches of the Carpathians in the northeast, the western parts of the Rhodope massif in the east, and the Dinaric Mountains along the Adriatic coast to the northwest and west [2], is classified within the Holarctic region and the Euro-Mediterranean subregion. Within this subregion, the native freshwater fish species are categorized into Central European and Southern European divisions. A significant portion of the fauna in Lake Ohrid is freshwater-based and can be traced back to the ancient saline waters of the Tertiary period that once existed in the Balkan region [3]. This fauna represents a direct continuation of the old Balkan freshwater species, which closely resembled the freshwater fauna found throughout Eurasia [4]. The preservation of Tertiary species in Lake Ohrid, as well as in numerous other water bodies in the Western Balkans, can be attributed to the fact that these waters have remained isolated from the Aegean and Danube, and thus from the stronger Pontic and Aegean faunal elements that adapted more readily to changing conditions. As noted by Ref. [3] "The archaic inhabitants of Lake Ohrid were doubly protected from extinction; not only did the long continuity of life in the lake contribute to this, but their isolation from external invasions also allowed for the survival of its relict fauna."

The fish population found in Lake Ohrid and the Adriatic basin is biogeographically linked to the Western Balkans' fauna, which is distinctly characterized by its composition and particularly by its high number of endemic species. This distinctiveness sharply contrasts with the fish fauna of the Eastern Balkans and the wider European region [3]. The absence of numerous fish species common in the Danube and other European waters suggests that the Western Balkans' fauna has been isolated from the rest of Europe for a considerable time. Furthermore, the fish fauna of the Western Balkans is comprised of many endemic species that are clearly differentiated from other European fish. "All endemic fish species found in today's Ohrid and Prespa fish fauna, which are restricted to the waters of the Dinaric region, represent remnants of the ancient fauna that once occupied these waters and have survived to the present day due to their great age, stable living conditions, continuous existence, and long-term geographical isolation" [3]. Lake Ohrid is often referred to as a "museum of living fossils" [4] due to the unique survival of its endemic species.

2. Material and methods

The species descriptions in this work are based on both our research findings and those of other researchers [1, 2, 5–9]. The meristic fish Formula (Eq. (1)) provided for various species include their meristic characteristics, which account for the last doubly branched ray in the dorsal (pinna dorsalis) and anal fins (pinna analis) as one unit. Only scales perforated by the lateral line canal (linea lateralis) are counted in the lateral line. Gill rakers are counted from the outer side of the first-gill arch on the left side of the head. For a limited number of specimens, the count of vertebrae is taken, with the last vertebra being the urostyle. The following meristic characteristics were evaluated: the number of scales in the lateral line (linea lateralis), above the lateral line (linea lateralis superior), below the lateral line (linea lateralis inferior), and the counts of hard and soft rays in the dorsal fin (pinna dorsalis), anal fin (pinna analis), ventral fin (pinna ventrales), pectoral fins (pinnae pectorales), and tail fin (pinna caudalis). In the formulas for some species, both spinal vertebrae and branchiospines are represented as follows:

$$\text{Meristic fish Formula}: \text{D III } (8); 12-13\ (14); \text{A III } 12-14; \text{P I } 16; \text{V I } 8; \\ \text{l.l.}7-9-56-57-5-6 \tag{1}$$

The list presented includes species confined to the waters within the borders of the Republic of North Macedonia, specifically those that are predominantly found within the Macedonian territory. Most species are located in three lake systems (Prespa, Ohrid, and Dojran Lake) shared with neighboring countries. Other species primarily inhabit the Adriatic and Ionian ecoregions, with some extending northward into the Lake Skadar basin in Albania and Montenegro. A few species from the Vardar Basin, including the Strumica River basin, are restricted to the southern part of the Balkans, particularly within the waters of Greece. These belong to Eco region 420 (Southeastern Adriatic basin, Prespa, and Ohrid lakes) and Eco region 422 (Vardar River Basin with Dojran Lake and Strumica River) (**Figure 1**) [10].

Fish species are categorized by the IUCN Red List [11–16] into nine groups based on criteria such as population decline rate, population size, geographic distribution area, and fragmentation degree. There is an emphasis on the acceptability of applying any criteria in the absence of high-quality data, including potential future threats, as long as these can be reasonably substantiated.

Extinct (EX)—The species is confirmed to be no longer extant.

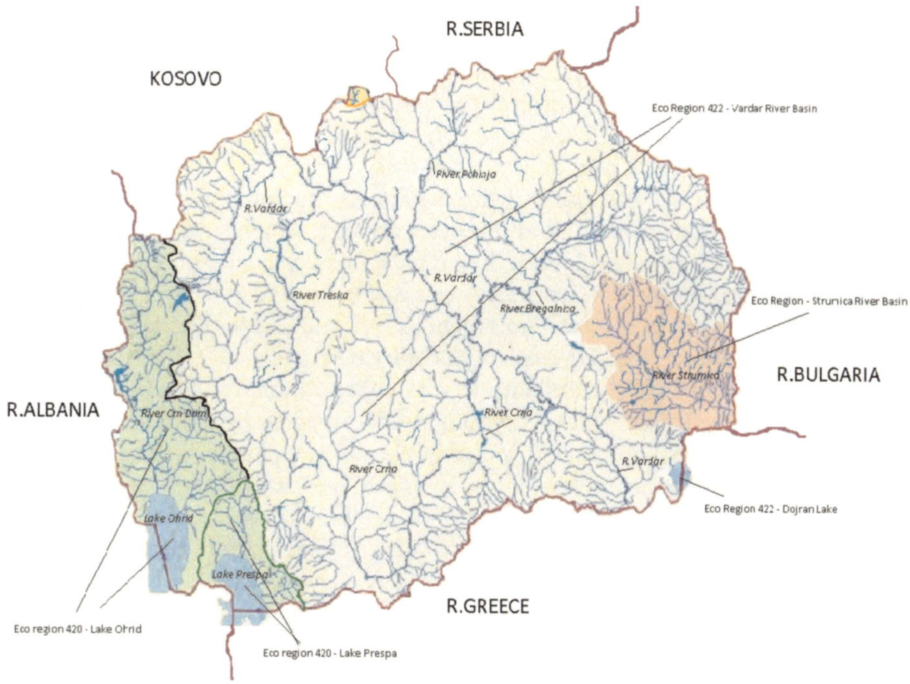

Figure 1.
Map of Eco region 420 Southeastern Adriatic basin (Lakes, Prespa, and Ohrid) and Eco region 422 (Vardar River Basin with Dojran Lake and Strumica River).

Extinct in the Wild (EW)—The species exists only in captivity, cultivation, and/or outside its native range, as presumed after thorough surveys.

Critically Endangered (CR)—The species is in an exceptionally critical state.

Endangered (EN)—There is a very high risk of extinction in the wild, meeting any of criteria A to E for Endangered.

Vulnerable (VU)—The species meets one of the five Red List criteria and is thus considered at high risk of unnatural (human-caused) extinction without further intervention.

Near Threatened (NT)—The species is close to being classified as endangered in the near future.

Conservation Dependent (CD)—This category has been removed, with its contents merged into Near Threatened; however, it may still exist for species not evaluated under current guidelines.

Least Concern (LC)—The species is unlikely to become endangered or extinct in the near future.

Data Deficient (DD)—The species has not been evaluated.

Not Evaluated (NE).

3. Results and discussion

Lake Ohrid is an ancient and unique freshwater ecosystem situated on the border between North Macedonia and Albania. It is recognized for its exceptional biodiversity and ecological importance. As one of Europe's oldest lakes, estimated to be over 2 million years old, it reaches a depth of around 286 m, which contributes to its distinct ecological features. The lake is particularly famous for its high level of endemism, supporting numerous species that are not found elsewhere globally, including various fish species, freshwater mollusks, and other aquatic organisms. The water in Lake Ohrid is noted for its clarity and low nutrient content, which are vital for sustaining its unique species. However, the delicate ecological balance is threatened by pollution from industrial, agricultural, and domestic sources.

The lake's ecosystem encompasses various habitats, including deep-water zones, littoral zones (shoreline areas), and underwater springs, which support diverse species and enhance the overall ecological diversity. Endemic species in Lake Ohrid include the Ohrid trout and various freshwater mollusks, which have adapted to the lake's stable, isolated environment over millennia. Key threats to the ecosystem consist of pollution, overfishing, invasive species, and climate change, all of which disrupt the ecological balance and pose risks to both water quality and native species.

Efforts to conserve Lake Ohrid's ecosystem involve designating it as a protected area and fostering international cooperation aimed at managing pollution, regulating fishing activities, and addressing the impacts of invasive species to safeguard the lake's unique biodiversity. Lake Ohrid's ecosystem is a rare and valuable natural resource, celebrated for its ancient origins and exceptional biodiversity. It is crucial to protect this fragile environment to maintain its ecological integrity and ensure the survival of its endemic species.

3.1 Endemic fish species in Lake Ohrid

3.1.1 Alburnoides ohridanus *(Karaman, 1928)*

Common Name: Ohrid spirlin, Ohridska gomnushka (Macedonian name)

$$\text{D III 8; A III 12; V II 7; P I 14;}$$
$$\text{l.l.8} - 43 - 46 - 3;\ \text{Vert.}38 - 40 \quad (2)$$

The Ohrid spirlin (*Alburnoides ohridanus*) is a small freshwater fish endemic to Lake Ohrid and also found in Lake Skadar. This species can be distinguished from the Vardar subspecies by its broader head, wider mouth, and shorter, blunter snout. A key feature, as noted by Ref. [9], is the presence of 11½ branched rays in the anal fin. The fish has a noticeable concavity at the nape, an upturned mouth, and 42-44 + 2-3 lateral line scales. The caudal peduncle is 1.6–1.8 times longer than its depth, with the snout length nearly equal to the eye diameter and interorbital distance. The Ohrid spirlin can reach a maximum size of about 90 mm in standard length, typically residing in the surf zones along the shores of Lake Ohrid. Its breeding season occurs from May to June. Adapted to freshwater environments in temperate climates, particularly within benthopelagic zones, the Ohrid spirlin is listed as Vulnerable (VU) on the IUCN Red List due to its limited distribution and specific habitat needs. It also appears in Annex III of the Bern Convention, which mandates specific conservation measures. While the species shows low to moderate sensitivity to hydropower projects, it faces moderate to high threats from dam constructions in the Balkan region.

3.1.2 Alburnus scoranza *(Bonaparte, 1845)*

Common Name: Ohrid scoranza, Ohridska plasica (Macedonian name)

$$\text{D III } 7 - 8;\ \text{A III } 12 - 15;\ \text{V II } 8;\ \text{P } 13 - 15;\ \text{Vert } 48 - 52 \quad (3)$$

Alburnus scoranza, commonly referred to as the Ohrid scoranza, is a small freshwater fish that is endemic to the central Western Balkans, specifically within the Drina watershed, which includes Lakes Skadar and Ohrid. This species thrives in larger and deeper river systems characterized by calm or slow-moving waters and can grow to a maximum length of 16 cm. The anal fin is situated beneath or just behind the last branched dorsal ray, while the ventral keel is exposed for roughly two-thirds of the distance from the anus to the base of the pelvis. The pectoral fin does not reach the base of the pelvis and is devoid of a dark lateral stripe. The Ohrid scoranza can attain a standard length of 158 mm and is distinguished by its broader head, wider mouth, and shorter, blunter snout.

The Ohrid scoranza is found in the basins of Lakes Skadar and Ohrid, along with their tributaries, and is also present in the Crn Drim River. While it predominantly inhabits lacustrine environments, it also occupies tributaries that feed into these lakes. Spawning occurs along the shores of the lakes or in shallow habitats of tributaries.

This species is noted for its social behavior, often forming large flocks and having a lifespan of up to 5 years. It typically spawns for the first time at 1 year of age, with spawning activities in Lake Ohrid taking place from May to June when water temperatures range from 18°C to 23°C. Eggs are deposited on sandy, gravelly, and rocky substrates at depths of approximately 0.3 m and hatch within about 4 days. The Ohrid scoranza primarily feeds on Cladocera and Copepods.

The IUCN Red List classifies the Ohrid scoranza as Least Concern (LC), suggesting that it is relatively plentiful but has a restricted geographical distribution. Ongoing monitoring is essential due to its limited range.

3.1.3 Barbatula sturanyi *(Steindachner, 1892)*

Common Name: Ohrid stone loach, Ohridski kamnar (Macedonian name)

$$\text{D IV 7; A III 5; P I 10} \tag{4}$$

Barbatula sturanyi is a small freshwater fish species primarily found in the Lake Ohrid basin and the upper Drin region, encompassing parts of North Macedonia and Albania. This species can be differentiated from other European Barbatula species by specific features: adult fish have fins covered with small tubercles, especially apparent during spring, giving the skin a slightly rough texture. The caudal fin is truncated, and the dorsal fins origin is slightly ahead of the pelvic fin's origin. The depth of the caudal peduncle is 1.6–2.1 times its depth and measures 1.7–2.0 times the body depth. This species can grow up to 100 mm in standard length.

Barbatula sturanyi typically inhabits springs along lake shores and streams with stony bottoms. Currently, there is no specific biological data available for this species.

In terms of conservation status, it is classified as Least Concern (LC), indicating no significant immediate threat to its population.

3.1.4 Barbus rebeli *(Koller, 1925)*

Common Name: Western Balkan barbell, Ohridska (crna) mrena (Macedonian name)

$$\begin{gathered}\text{D IV 8; A III 5; V II 8; P I } 15-16;\\ \text{l.l.}11-12-49-53-7-8\end{gathered} \tag{5}$$

Barbus rebeli, commonly referred to as the Western Balkan barbel, is a freshwater fish species native to the Adriatic basin, particularly found in the Drin River drainage system, including Lakes Skadar and Ohrid, spanning across Montenegro, Albania, and North Macedonia. This species is easily recognized by its distinctive physical characteristics, with the head, body, and fins dotted with numerous small black spots that are more prominent and darker on the back. The lower lip features a small median lobe, and the simple pelvic fin ray is approximately equal in length to the second branched ray. The posterior edge of the dorsal fin is either straight or slightly concave. Barbus rebeli can grow up to 250 mm in standard length.

In terms of habitat, *Barbus rebeli* thrives in both lake and river environments. In rivers and streams, it prefers clean water bodies with vegetation. Despite being widespread and locally abundant, the species faces significant conservation challenges. Many small streams within its range dry up during the summer, leading to fragmented populations, while overfishing is also a concern.

Currently, *Barbus rebeli* is classified as Least Concern (LC) on the IUCN Red List but is also listed under Annex V of the EU Habitats Directive and is protected by the Bern Convention.

3.1.5 Cobitis ohridana *(Karaman, 1928)*

Common Name: Ohrid spined loach, Ohridska shtipalka (Macedonian name)

$$\text{D III } 6-7\text{; A III 5; V II 6; P 18; C 16} \tag{6}$$

Cobitis ohridana, commonly referred to as the Ohrid spined loach, is a small fish species that is exclusive to the Southern Balkans, primarily located in Lake Ohrid, the Drim River, and its associated tributaries. This species is notable for its relatively higher and blunter head compared to other varieties, as well as its mottled appearance, which is characterized by prominent brown spots across its body. It displays a distinctively striped back and a heavily marked head, alongside having weaker fin rays in comparison to other species within the same genus. Female Ohrid spined loaches can attain a standard length of at least 65 mm, while males tend to be slightly smaller.

These fish are typically found in rivers and lakes, often dwelling over fine to muddy sand substrates and among algae.

Regarding its conservation status, *Cobitis ohridana* is listed as Least Concern (LC) on the IUCN Red List. Although there are stable populations in certain regions, the species' habitats face threats from local human activities, which restrict its distribution.

3.1.6 Gobio ohridanus *(Karaman, 1924)*

Common Name: Ohrid gudgeon, ohridski mrenec (Macedonian name)

$$\begin{gathered} \text{D III 7; A III 6; V II 7; P I 15; C 19;} \\ \text{l.l.}6-39-40-3;\text{Vert.}38 \end{gathered} \tag{7}$$

Gobio ohridanus is a small fish species that is native to Lake Ohrid, it is present in R.N. Macedonia and Albania. Its notable characteristics include a head length that comprises 29–34% of its standard length (SL) and the presence of 4–5 scales in the area between the anus and the origin of the anal fin. The species is devoid of scales on the chest region between the pectoral fins and possesses between 36-39 + 2-3 lateral line scales. The diameter of the eye ranges from 1.1 to 1.5 times the distance between the eyes, while the length of the snout makes up 40–45% of the head length (HL). Importantly, individuals exceedingly approximately 70 mm SL display a noticeably concave area between the eyes. This fish also exhibits 7–10 irregular mid-lateral blotches and smaller black spots on its back, in addition to a series of black subdistal spots along the rays of the dorsal fin. *Gobio ohridanus* can reach a maximum size of 130 mm SL and typically resides in the lake's sandy to pebble substrate. It tends to be social and breeds during the months of June and July.

This species is listed as Vulnerable (VU) on the IUCN Red List due to its limited distribution, which makes it prone to environmental shifts.

3.1.7 Pachychilon pictum *(Heckel and Kner, 1858)*

Common Names: Albanian roach; Moranec (Macedonian name)

$$\begin{gathered} \text{D III }7-9;\text{ A III }8-10;\text{ V II 7; P I }11-14; \\ \text{l.l.}7-9-40-47-2-4;\text{Vert.}34-38;\text{Branchiospines }9-14 \end{gathered} \tag{8}$$

Pachychilon pictum, commonly known as the Moranec, is a freshwater fish species that is native to the Southern Balkans, particularly within the Adriatic basin, encompassing the drainages of the Drin River. This fish is characterized by its body, which is adorned with numerous dark brown spots that vary in shape and size. Young individuals typically exhibit a prominent dark stripe along their bodies, although this characteristic may diminish in preserved samples. The lower lip of *Pachychilon pictum* is significantly thick and features a long median lobe. This species can attain a maximum length of approximately 18 cm, with some individuals reaching lengths of up to 20 cm and weights of around 70 g.

Pachychilon pictum prefers habitats such as slow-moving rivers, canals, backwaters, and lakes, usually found near the shorelines. It is a social species that often forms large flocks. The spawning period occurs between April and June, typically over gravel beds or amidst vegetation when water temperatures rise above 12°C. The eggs, which are adhesive, adhere to the substrate and have an incubation duration of about 210 hours at temperatures ranging from 12.0°C to 17.5°C.

From a conservation perspective, *Pachychilon pictum* is classified as Least Concern (LC) on the IUCN Red List. Although it is relatively abundant within its native range, its localized distribution and specific habitat requirements render it susceptible to environmental changes and human activities. This species plays a vital role in the local aquatic biodiversity.

3.1.8 Pelasgus minutus *(Karaman, 1924)*

Common Name: Ohrid minnow, Ohridsko grunče (Macedonian name)

$$\text{D III 7; A III 6} - \text{7; V (I) II (III) 6} - \text{8; P I 13} - \text{15;} \tag{9}$$

Pelasgus minutus is characterized by a unique body structure that features a limited number of scales. It is distinguishable from *P. epiroticus* and the fish found in Prespa Lake due to its longer body and shorter tail, as well as its larger eyes and broader, elongated head. The lower lip extends slightly beyond the upper lip, and although the mouth opening is relatively small, it is wider than that of *P. epiroticus*. The scales are comparatively larger, with the largest scale diameter on a fish measuring 50 mm reaching 1 mm, which is three times the size of the scales found on *P. epiroticus* of the same length. This species can reach a maximum length of 6 cm.

From a conservation perspective, *Pelasgus minutus* is classified as on the IUCN Red List Data Deficient (DD).

Its natural habitat is the freshwater littoral zone in lakes overgrown with vegetation.

3.1.9 Rutilus ohridanus *(Karaman, 1924)*

Common Name: Ohrid roach, ohridski grunec (Macedonian name)

$$\begin{gathered}\text{D III 10} - \text{11; A III 11; V II 8; P I 16} - \text{17; C 19;}\\ \text{l.l.9} - 44 - \text{4; Vert.42}\end{gathered} \tag{10}$$

Rutilus ohridanus is a freshwater fish species indigenous to the Balkan Peninsula, predominantly located in Lakes Ohrid and Skadar. It is present in North Macedonia, Albania, and Montenegro. This fish can be distinguished from other Rutilus species in the area by several distinct characteristics, such as its translucent to grayish fins and a rounded snout, which forms a 15–25° angle with the horizontal line at the mouth. The lower part of its head features a smooth, rounded contour at the lower jaw joint, while the body exhibits a silvery appearance when alive. *Rutilus ohridanus* can reach a maximum length of approximately 130 mm and is typically found in lake environments, particularly along the shores, where it spawns on gravel, stones, and inundated regions. It also inhabits warmer rivers and streams, especially in areas with little to no current.

The lifespan of *Rutilus ohridanus* can last up to 7 years, with its spawning season occurring between April and May. The sticky yellow-orange eggs hatch in about 9 days at temperatures ranging from 13°C to 17°C, and the fish primarily feeds on plankton and plant matter.

Although facing various environmental pressures, *Rutilus ohridanus* is currently designated as Least Concern (LC) on the IUCN Red List.

3.1.10 Salmo aphelios *(Kottelat, 1997)*

Common Names: Summer trout, Letnica (Macedonian name)

$$\begin{gathered} \text{D } 10-14 \text{ A } 9-12 \text{ P } 12-15 \text{ V } 8 \\ \text{l.l.} 80-100 \text{ Vertebrae } 30-34 \end{gathered} \tag{11}$$

Salmo aphelios, often referred to as the summer trout or letnica is a trout species indigenous to the eastern coastline of Lake Ohrid in North Macedonia. This species can reach a minimum standard length (SL) of 400 mm and is notably recognized for its unique orange flesh, distinguishing it from other trout species found in the lake. *Salmo aphelios* thrives in lacustrine habitats and typically breeds from May to July in proximity to sub-lacustrine springs along the eastern lake shore.

Currently, its conservation status is categorized as Data Deficient (DD), highlighting the need for further research to gain a comprehensive understanding of its population dynamics and potential threats.

3.1.11 Salmo balcanicus *(Karaman, 1926)*

Common Names: Struga trout, Kresnica (Macedonian name)

$$\begin{gathered} \text{D IV } 9;\text{A IV } 7-8;\text{V II } 8;\text{P I } 11-13 \text{ C } 19 \\ \text{l.l.} 21-27-115-122-21-23 \text{ Ap.pyl.} 49-74 \end{gathered} \tag{12}$$

Salmo balcanicus, commonly known as the Struga trout or Kresnica, was first described by Karaman in 1926. This species can grow to a minimum standard length of 400 mm and is notable for its pinkish flesh. It is identified by specific features: dorsal fins with nine rays, anal fins with 7–8 rays, and pectoral fins with 11–13 rays. *Salmo balcanicus* inhabits the northwestern shore of Lake Ohrid, which spans both Albania and North Macedonia, as well as the lake's outlet. It predominantly resides in lacustrine habitats but migrates to the outlet for spawning, which takes place from October to January, especially at this outlet.

The conservation status of *Salmo balcanicus* is currently classified as Data Deficient (DD). This species is potentially critically endangered due to habitat degradation, as its original spawning locations have become inaccessible due to the construction of a weir and two reservoirs blocking the lake's outlet.

3.1.12 Salmo letnica *(Karaman, 1924)*

Common Name: Ohrid trout, ohridska pastrmka (Macedonian name)

$$\text{D IV } 8-9, \text{A IV } 8-9; \text{V II } 8; \text{P I } 11, \text{C}19$$
$$\text{l.l.}15-17-105-13-15 \text{ Vertebrae } 53-57 \quad (13)$$

Salmo letnica, commonly referred to as the Pestani trout, is a trout species that is native to the eastern shoreline of Lake Ohrid in North Macedonia. This species is significant due to its freshwater habitat and can grow to a standard length (SL) of up to 760 mm. Adult Pestani trout are usually silvery in color, featuring black spots that are predominantly found on the upper section of their bodies. If present, red spots are typically located along the lateral line. The flesh of *Salmo letnica* is characterized by a pinkish tint. This species is primarily found in the littoral and sublittoral zones of Lake Ohrid.

Pestani trout achieve sexual maturity between the ages of 3 and 4 years, with a maturity length (Lm) that generally falls between 36 and 39.7 cm. The spawning period occurs from January to February. The dietary habits of *Salmo letnica* change as they mature; juvenile trout mainly consume zooplankton, whereas adults diversify their diet to include both zooplankton and smaller fish, with *Alburnus scoranza* being a key component.

Although *Salmo letnica* plays a vital role in the ecosystem of Lake Ohrid, it is currently classified as Data Deficient (DD) by conservation organizations, which suggests that additional research is necessary to accurately evaluate its population status and identify potential threats.

3.1.13 Salmo lumi *(Poljakov, Filipi, Basho and Hysenaj, 1958)*

Common Name: Lumi trout, recna pastrmka (Macedonian name).

Salmo lumi, commonly referred to as the lumi trout, is a trout species that inhabits the northern and western banks of Lake Ohrid, which is located in both Albania and North Macedonia. This trout can reach a minimum standard length of 380 mm and primarily resides in lacustrine habitats, where it breeds in the lake's tributaries and estuaries. The distinguishing characteristics of *Salmo lumi* remain ambiguous, and its existence is somewhat questionable due to insufficient conclusive evidence. The breeding season for this species occurs from December to February.

Currently, the conservation status of *Salmo lumi* is categorized as Data Deficient (DD), highlighting the necessity for further studies to confirm its presence and evaluate its population status. Despite the existing uncertainties, *Salmo lumi* is provisionally acknowledged as a valid species pending additional information.

3.1.14 Salmo ohridanus *(Steindachner, 1892)*

Common Name: Ohrid belvica, Ohridska belvica (Macedonian name)

$$\text{D IV } 8-9; \text{A IV } 8-9; \text{V II } 8; \text{P I } 11 \text{ C } 19;$$
$$\text{l.l.}15-17-100-114-13-15 \quad (14)$$

Salmo ohridanus, widely referred to as belvica, is a trout species indigenous to Lake Ohrid, and it is present in Albania and North Macedonia. This fish is recognized by specific traits, including 100–114 lateral line scales and 11–12 scale rows situated between the lateral line and the adipose fin. Unlike other salmon species found in the Balkan Peninsula, *Salmo ohridanus* generally has 54–55 vertebrae. Its snout is particularly blunt, and its length is approximately equal to the diameter of its eye. The fish's body typically exhibits a yellowish-silver hue, occasionally adorned with small, x-shaped pinkish spots.

Salmo ohridanus can reach a maximum standard length of approximately 350 mm. This freshwater species resides in Lake Ohrid, occupying depths from the surface down to about 20 m during the summer months, and tends to move closer to the shore in winter. The spawning period occurs between December and February along the lake's perimeter, with eggs incubating for around 45–55 days at a temperature of 10°C. This species is known to form flocks.

Despite its resilience, *Salmo ohridanus* is currently listed as Vulnerable (VU) under criterion D2, largely due to the persistent threat of intentional hybridization that has affected it for the last 50 years. Historically, this species was classified under a separate genus, *Acantholingua*, but has since been reclassified within the genus *Salmo*.

3.2 Endemic fish species in Lake Prespa

Lake Prespa, located at the intersection of North Macedonia, Albania, and Greece, is a notable freshwater ecosystem. It is classified as a large, shallow lake, with a maximum depth of approximately 54 m. This lake is part of a transboundary system that encompasses both the larger and smaller Prespa Lakes, collectively referred to as the Prespa Basin. The lake harbors a diverse array of flora and fauna, featuring numerous endemic species. It is particularly recognized for its rich bird populations, which include pelicans, herons, and various waterfowl. The fish community within Lake Prespa includes unique species such as the Prespa trout and the Prespa carp.

The lake offers a variety of habitats, including shallow waters, reed beds, and adjacent wetlands, all of which are essential for the breeding and feeding activities of various species. Historically, Lake Prespa has been noted for its clear waters; however, it currently faces significant challenges, including pollution and eutrophication due to agricultural runoff and wastewater discharge. Among the unique inhabitants of the lake are the endemic Prespa trout (*Salmo prespensis*) and several distinctive cyprinid species, each adapted to the lake's specific environmental conditions.

The ecosystem of Lake Prespa is under threat from pollution originating from agricultural and industrial activities, the presence of invasive species, habitat destruction, and fluctuations in water levels, all of which can deteriorate water quality and disrupt the ecological equilibrium. Conservation initiatives are underway to safeguard the unique biodiversity of the lake and tackle pollution concerns. These efforts encompass habitat restoration, pollution management, and international collaboration among the nations that share the lake. The ecosystem's rich biodiversity and ecological importance render Lake Prespa a critical area for conservation and environmental stewardship.

3.2.1 Alburnoides prespensis *(Karaman, 1924)*

Common Name: Prespa spirlin, prespanska gomnushka (Macedonian name)

$$D\ III\ 8; A\ III\ 10-12; V\ II\ 7; P\ I\ 13-14; C\ 19$$

$$\text{l.l.}9 - 44 - 48 \text{ vert } 38 - 40 \tag{15}$$

Alburnoides prespensis, identified by Karaman in 1924, is a species that is exclusively native to Lake Prespa and is present across Greece, North Macedonia, and Albania. Key characteristics of *Alburnoides prespensis* include relatively smaller eyes when compared to *A. a. bipunctatus*, with the eyes measuring about four times the length of its head. Additionally, it possesses a more prominent dorsal fin and has fewer branched rays in its anal fin.

This fish primarily resides in the Southern Balkans, especially within Lake Prespa and its surrounding tributaries. It is usually located near the shorelines of the lake and often swims in flocks.

The species is found in the surf zone along the lake's edges and typically spawns between May and June, attaining a maximum length of around 9 cm.

In terms of conservation status this species is currently categorized as Vulnerable (VU) D2 ver 3.1 in the IUCN Red List and listed in Annex III of the Bern Convention. *Alburnoides prespensis* faces threats from water extraction, pollution, and the effects of invasive species, leading to a decline in its population.

3.2.2 Alburnus belvica *(Karaman, 1924)*

Common Name: Prespa bleak, Belvica, Cironka (Macedonian name)

$$\begin{gathered}\text{D III } 8;\text{ A III } 12 - 14;\text{ V II } 8;\text{ P I } 15 - 17;\text{ C } 19\\ \text{l.l.}10 - 50 - 55 - 3\end{gathered} \tag{16}$$

Alburnus belvica, another endemic species is confined to Lakes Megali and Mikri Prespa (in northwestern Greece, Albania, and North Macedonia), as well as their connected tributaries.

This freshwater fish displays a streamlined body typical of the *Alburnus* genus, with a terminal mouth and relatively small, silver-colored scales. It can grow up to 20 cm in total length.

Preferring open water, *Alburnus belvica* is often found along shorelines and exhibits grouping behavior in oxygen-rich environments. This fish species migrates to tributaries or shallow lake areas for spawning during the spring and feeds primarily on zooplankton and small invertebrates.

The IUCN Red List classifies it as Vulnerable (VU) D2 ver 3.1., due to habitat degradation, pollution, water abstraction, and the influx of nonnative species, which have contributed to the decline of its population. Key threats include habitat loss from water extraction for agriculture, pollution from agricultural runoff, and competition or predation by introduced species. Conservation measures focus on population monitoring, habitat protection, and pollution mitigation.

3.2.3 Barbus prespensis *(Karaman, 1924)*

Common Name: Prespa barbell, Prespanska mrena (Macedonian name)

$$\begin{gathered}\text{D IV } 8;\text{ A III } 5;\text{ V II } 8;\text{ P I } 16 - 17;\text{ C}19\\ \text{l.l.}12 - 14 - 52 - 64 - 10 - 11 \text{ vert.}41 - \text{sp.br } 8 - 9\end{gathered} \tag{17}$$

Barbus prespensis a species native to lakes Megali and Mikri Prespa present in northwestern Greece, Albania, and North Macedonia along with connecting rivers and tributaries. This freshwater fish is characterized by a solid, elongated body typical of the genus Barbus, and a slightly flattened head with whiskers located near its mouth. Its color generally varies from olive-brown on the back to lighter shades of silver on the sides and belly, with individuals usually reaching a length of 25–30 cm.

This species prefers clear, fast-moving water, streams and areas with aquatic vegetation, and often lives near the bottom.

This fish is classified as Least Concern (LC) under version 3.1 of the Conservation Status Assessments in the IUCN Red List.

Generally, these fish feed on invertebrates, small crustaceans, and plant matter. Migrates to gravel beds in tributaries or shallow areas of the lake to spawn during spring.

The key threats to this species are pollution from agricultural runoff, habitat alteration due to water extraction, and competition or predation by invasive species.

3.2.4 Chondrostoma prespense *(Karaman, 1924)*

Common Name: Prespa nase, Prespanski skobust (Macedonian name)

$$\begin{gathered} \text{D III 8; A III 9; V II 8; P I } 16-17\text{; C } 19 \\ \text{l.l.} 10-58-60-5\text{; vert } 44 \end{gathered} \tag{18}$$

Chondrostoma prespense is an endemic species found in the Prespa Lake basin with its tributaries, present in Albania, Greece, and North Macedonia. This small to medium-sized freshwater fish is characterized by an elongated body, a slightly rounded back, and a subterminal mouth with a pronounced keratinized lower lip. The color ranges from grayish to silver on the sides, with a darker on back and can reach a length of up to 20 cm.

Prespa nase prefers clear, flowing freshwater environments. It usually inhabits shallow vegetated areas along lake shores and is known to be a benthic feeder, consuming algae, and detritus.

Due to its limited distribution and the multitude of threats it faces, this species is classified as Vulnerable (VU) D2 ver 3.1. from the IUCN Red List. Key threats include habitat degradation, pollution and water extraction, along with competition from nonnative species. Conservation measures focus on protecting habitats, improving water quality, and preventing the introduction of invasive species.

3.2.5 Cobitis meridionalis *(Karaman, 1924)*

Common Name: Prespa loach, Prespanski mrenec (Macedonian name)

$$\text{D III } 6-7\text{; A IIII 5; V II 6P I 8; C 16} \tag{19}$$

Cobitis meridionalis is an endemic species of the Balkans, a species unique to the Prespa Lake basin, present in Albania, Greece, and North Macedonia (in the LJIlake and in the tributaries leading to Prespa Lake). This small and slender freshwater fish typically grows to about 10 cm in length, features dark brown to black spots set against a lighter, sandy-hued body, and possesses three pairs of whiskers around its mouth.

It prefers clear, slow-moving freshwater habitats with sandy or muddy bottoms, often burrowing into the substrate for camouflage and protection. This species is primarily nocturnal and feeds on benthic invertebrates, and the spawning period is from late spring to early summer.

Cobitis meridionalis is recognized as Vulnerable (VU) D2 ver 3.1. in the IUCN Red List, facing various threats that threaten its long-term survival.

The main threats include habitat degradation from pollution from agricultural runoff and industrial activities, water abstraction (over-abstraction for irrigation and human consumption), and the introduction of invasive species which increases competition for resources and increases the risks of predation.

Conservation efforts aim to protect habitats, mitigate pollution, and create protected areas within their natural distribution range.

3.2.6 Economidichthys pygmaeus *(Holly, 1929)*

Common Name: Western Greece goby,

Economidichthis pigmaeus is a small benthic fish, typically reaching a maximum length of 6 cm. This species is mainly found in western Greece, particularly within the Ionian region, where it resides in lowland rivers, springs, and marshes, including areas significantly altered by human activity, such as canals. In our study of Lake Prespa [16, 17], we observed the presence of *Economidichthis pygmaeus*; however, we have not yet established whether it is an introduced species or a native (endemic) fish in that ecosystem.

It is currently classified as Least Concern (LC) on the IUCN Red List.

3.2.7 Pelasgus prespensis *(Karaman, 1924)*

Common Name: Prespa minnow, Prespansko malo grunche (Macedonian name)

$$D\ III\ 7;\ A\ III\ 6-7;\ V\ (I)\ II\ (III)\ 6-8;\ P\ I\ 13-15; \qquad (20)$$

The Prespa minnow, a species native to the Prespa Lakes, is present in Greece, North Macedonia, and Albania. This small freshwater fish is characterized by its elongated body and rounded shape, often displaying a silvery or grayish hue with a darker dorsal surface and subtle markings. It can reach a length of up to 6–8 cm.

Exclusive to the Prespa Lake basin, *Pelasgus prespensis* is present in shallow, vegetated areas along the edges of lakes, as well as in slow rivers and streams. It favors clean, oxygen-rich waters, usually found in areas with gravel or sandy bottoms.

The Prespa minnow is omnivorous, consuming small invertebrates, algae and detritus, and is known to exhibit schooling behavior, with spawning occurring in spring and early summer.

It is currently classified as Endangered (EN) B1ab (iii, iv, v) + 2ab (iii, iv, v) version 3.1. on the IUCN Red List. This species is very sensitive to changes in its habitat.

The primary threats to its survival are habitat degradation (pollution resulting from agricultural practices and urban expansion), the abstraction of water for irrigation and domestic purposes, and the introduction of nonnative species that create competition.

Conservation efforts are aimed at protecting habitats, controlling pollution, and ensuring natural water levels are maintained. Ongoing population monitoring and measures to prevent habitat loss are key to the conservation of this species.

3.2.8 Rutilus prespensis *(Karaman, 1924)*

Common Name: Prespa roach, Prespanski grunec, Prespanska pisa (Macedonian name)

$$\text{D III 9; A III 8; V II 8; P I 16; C 19}$$
$$\text{l.l.}8 - 38 - 40 - 3 \text{ Vert } 38 \quad (21)$$

The Prespa roach is a species of fish born (unique) in the basin of the Prespa Lake, primarily living in the Prespa Lake and occasionally found in its tributaries (present in Greece, North Macedonia, and Albania). Scientifically known as *Rutilus prespensis*, it is a medium-sized freshwater fish that exhibits the laterally compressed body shape typical of its genus. The fish is silver-gray in color with a darker olive back and usually reaches a length of 25–30 cm, although most specimens are smaller.

It prefers open water but can also inhabit areas with dense aquatic vegetation along the coast. *Rutilus prespensis* is found in clean, oxygen-rich environments with sandy, gravelly, or muddy substrates.

As an omnivore, this fish feeds on a variety of food sources, including algae, detritus, small invertebrates, and plankton. Its spawning occurs in the spring, where it lays eggs on submerged vegetation or substrates in shallow water and is known to form flocks, especially during the juvenile phase.

Listed as Vulnerable (VU) according to D2 ver 3.1 criteria on the IUCN Red List, the Prespa roach faces significant threats due to its limited distribution.

Key threats include habitat degradation from agricultural runoff and urban pollution, water abstraction for agricultural and domestic purposes, and competition or predation by invasive species.

Conservation initiatives aim to restore and protect habitat, improve water quality, and manage invasive species. Regular monitoring of population trends and protection of their limited range is vital to the ongoing survival of the species.

3.2.9 Salmo peristericus *(Karaman, 1938)*

Common Name: Prespa trout or Peristeri trout, Pelisterska pastrmka (Macedonian name).

Salmo peristericus, a species native to the Prespa Lake basin, which covers Greece, North Macedonia, and Albania. This variety of trout shows the slender, elongated figure typical of the genus *Salmo*, characterized by a predominantly silver body with a darker olive-brown dorsal side, decorated with small black spots and occasionally with red or orange spots along its fins. Its length can range from 25 to 40 cm, depending on environmental conditions.

Endemic to the southwestern Balkans, *Salmo peristericus* is found in the waters of the Perister (Baba) mountain range. It mostly lives in cold mountain streams that flow fast and flow into Lake Prespa, favoring rocky substrates and deep pools for refuge. This species is particularly well-adapted to the lower temperatures that prevail at higher altitudes.

As a primarily carnivorous fish, the Prespa trout consumes aquatic invertebrates, small fish, and sometimes terrestrial insects. The spawning season is from late fall to early winter in the gravel beds of fast-flowing streams, with the species exhibiting territorial behavior.

This species is classified as Endangered (EN) B1ab(iii) + 2ab(iii, iv, v) according to version 3.1.

Threats include habitat degradation (activities such as deforestation, agricultural runoff, and urban development) that negatively affect water quality and spawning areas, water abstraction (the use of water for irrigation and consumption by humans alters stream flows, which negatively affects habitats) and the introduction of nonnative trout species poses risks through hybridization and competition for resources.

Conservation measures are focused on protecting and restoring habitats, regulating water use, and managing invasive species. In addition, artificial breeding initiatives have been launched to help the population recover.

3.2.10 Squalius prespensis *(Fowler, 1977)*

Common Name: Prespanski klen (Macedonian name)

D III 8; A III 8; V II 8; P I 15 – 17; C 19

l.l.7 – 43 – 44 – 3 vert 42 (22)

Squalius prespensis is a species native to the Prespa Lake basin, present in Greece, North Macedonia, and Albania. This freshwater fish belongs to the Cyprinidae family and can grow up to 25 cm in total length. Originally, it was classified as a subspecies of European chub. Currently, *Squalius prespensis* is widespread within its limited range and appears to have experienced a recent population increase.

This species is categorized as Least Concern (LC) according to version 3.1. from IUCN Red List.

However, it faces potential threats from factors such as overfishing, water extraction, and environmental pollution.

3.3 Endemic fish species in the Aegean water area of North Macedonia (Lake Dojran, River Vardar, and River Strumica)

The Aegean water area in North Macedonia encompasses primarily the Vardar River, along with Lake Dojran and the Strumica River, which forms an independent river ecosystem that ultimately flows into the Aegean Sea.

Dojran Lake

Dojran Lake is a small, yet historically significant lake located on the border between North Macedonia and Greece. It is the smallest of the three major lakes in North Macedonia, along with Prespa and Ohrid.

The lake spans an area of approximately 43.1 km^2, with 27.3 km^2 attributed to North Macedonia and 15.8 km^2 to Greece. With a maximum depth of around 10 m, it is relatively shallow compared to other significant lakes in the region.

Dojran Lake is renowned for its rich biodiversity, particularly regarding fish species, and has historically been important for traditional fishing practices, including the use of birds like cormorants to assist in catching fish.

Over time, the lake has faced ecological challenges, including pollution, overfishing, and changes in water levels due to climate change and water extraction. Conservation projects have been initiated to preserve the lake's ecosystem.

Recent years have seen increased efforts to restore and maintain the ecological balance of the lake. International cooperation, especially between North Macedonia and Greece, has played a vital role in these conservation efforts. Sustainable tourism practices are being promoted to ensure that the lake remains a viable natural resource for future generations.

Vardar River

The Vardar River is one of the most significant rivers in the Balkan Peninsula, traversing North Macedonia and Greece. It plays a crucial role in the region's ecology, economy, and culture.

Stretching approximately 388 km (241 miles), it is the longest river in North Macedonia, originating near the village of Vrutok, close to the town of Gostivar in the Sharr Mountain range.

Flowing southeast through North Macedonia, the Vardar passes through major cities, including Skopje (the capital) and Veles, before entering Greece, where it is referred to as the Axios River, continuing its course through the Greek region of Macedonia until it empties into the Aegean Sea near Thessaloniki.

The Vardar River basin covers around 25,000 km^2, predominantly in North Macedonia, with a smaller portion in Greece. This river supports diverse ecosystems, providing habitats for numerous fish, birds, and other wildlife that rely on its waters.

However, pollution from industrial activities, agricultural runoff, and urban waste has adversely affected water quality in certain sections of the river. Ongoing efforts aim to improve water management and reduce pollution, particularly in urban areas like Skopje.

Challenges faced by the Vardar River include pollution, water management issues, and the impacts of climate change. Cross-border cooperation between North Macedonia and Greece is essential for effective conservation and sustainable resource utilization. Initiatives aimed at enhancing water quality, preserving biodiversity, and promoting sustainable development in the river basin are currently in progress, supported by both national governments and international organizations.

Strumica River

The Strumica River is a significant watercourse located in the southeastern part of North Macedonia and extending into Bulgaria and Greece. It plays a vital role in local geography, ecology, and the livelihoods of nearby inhabitants.

Approximately 114 km long, the Strumica River originates from the slopes of the Belasica mountain range in Macedonia. It flows southwest into North Macedonia, passing through the town of Strumica, from which it derives its name, before continuing southward to join the Struma River (known as the Strymon River in Greece) near the North Macedonia-Greece border.

The Strumica River basin covers an area of around 1649 km^2, primarily within North Macedonia, with smaller portions in Bulgaria and Greece.

This river supports a variety of ecosystems, providing habitats for numerous fish, birds, and other wildlife. The valley of the Strumica River is rich in flora and fauna, contributing significantly to the region's biodiversity. However, the river faces environmental challenges, including pollution from agricultural runoff, industrial waste, and untreated sewage. Deforestation and soil erosion in surrounding areas exacerbate sedimentation issues.

Efforts to protect the Strumica River include initiatives to reduce pollution, improve water management, and promote sustainable agricultural practices. Environmental organizations and local authorities collaborate to address these challenges, aiming to preserve the river's ecological health and ensure it remains a vital resource for future generations.

3.3.1 Alburnus macedonicus *(Karaman, 1928)*

Common Name: Belovica, Dojranska plashica (Macedonian name)

$$\begin{gathered} \text{D III 8; A III 15; V II 8; P I } 13-15, \\ 1.1.8-9-50-52-3 \end{gathered} \tag{23}$$

Alburnus macedonicus is endemic to the Dojran Lake (present in Greece and R.N. Macedonia) as well as in the Vardar and Strumica rivers.

Dojranska plashica (belovica) shows notable morphological differences compared to the typical Alburnus form, especially with the dorsal fin being less prominent. Some rays are shorter, leading to a reduced overall fin appearance. The dorsal fin usually contains about 15 branched rays, as opposed to the 18 of the standard variant, resulting in a shorter base. The length of the fin is approximately equal to the distance from the back of the operculum to the front edge of the eye, while in the Danube variant, this distance extends to the nostrils or beyond. Additionally, this variant has a slightly elongated head and mouth, along with a relatively lower body height. It can grow up to 15.8 cm. Belovica from Dojran Lake shows differences in nutrition (diet) throughout the year, with diet depending on habitat. During the winter, when it inhabits extensive areas of reeds, it feeds on algae, macrophytic plants, seeds of higher vegetation, and a variety of animal organisms, including gammarids, azeluses, oligochaetes, and larval and adult insects. In contrast, in pelagic waters, it feeds primarily on zooplankton, especially cladocerans, along with significant amounts of dipteran nymphs and adults.

This species are spawns along the shore of the lake Dojran. Its population is declining.

This species is categorized as Critically (CR) according to version 3.1. from IUCN Red List.

The lake's water level has dropped by 6 m since 1990 due to over-irrigation, posing a significant threat to its survival.

3.3.2 Pachychilon macedonicum *(Steindachner, 1892)*

Common Name: Macedonian roach, Mergur (Macedonian name)

$$\begin{gathered} \text{D III } 10-11\text{; A III } 10-11\text{, V II 8; P I } 16-17; \\ 1.1\,9-44-4 \end{gathered} \tag{24}$$

This species, part of the Cyprinidae family, is noteworthy for its limited distribution and ecological role in its native regions. Endemic to the Balkan Peninsula, primarily found in freshwater systems in North Macedonia and adjacent areas such as Greece and Albania. It is present in rivers, lakes, and streams with clear, slow-moving waters, especially within the Vardar River basin in North Macedonia.

Typically, *Pachychilon macedonicum* measures between 10 and 15 cm in length, with an elongated body with a silvery hue and prominent scales, reflecting a streamlined shape typical of many cyprinids.

Macedonian roach is an omnivorous feeder, it consumes a variety of small invertebrates, plant material, and detritus. Spawning occurs in the spring, typically in shallow waters. Juveniles develop in sheltered areas rich in vegetation.

This species is categorized as Data Deficient (DD) according to version 3.1. in IUCN Red List.

This species is at risk due to its limited habitat and sensitivity to environmental changes, including habitat degradation, pollution, and water flow alterations. Major threats consist of habitat destruction, agricultural and industrial pollution, water extraction, and competition from nonnative species.

Conservation measures involve protecting natural habitats, monitoring populations, and developing captive breeding programs if necessary.

As a native species, *Pachychilon macedonicum* supports the ecological balance of its freshwater habitats and contributes to local biodiversity. Research is ongoing to understand its biology and conservation needs, emphasizing the importance of protecting this species for the ecological health of the regions it inhabits.

3.3.3 Gobio bulgaricus *(Drensky, 1926)*

Common Name: Bulgarian gudgeon, Bugarski mrenec (Macedonian name).

Gobio bulgaricus, the Bulgarian gudgeon or Bulgarian spined loach, is a small fish of the Cyprinidae family. The first description in scientific literature was made in the early twenty-first century, and this species is a relatively new entry into ichthyological records. Although *Gobio bulgaricus* was only classified recently it is of great scientific interest because of its unique adaptations and ecological importance.

Populations have also been recorded in Serbia, Macedonia, and Greece. Even from a small portion of extinction, it reflects that freshwater ecosystems are connected, which supports specific habitat requirements for the said species. *Gobio bulgaricus* occurs in freshwater contexts, often inhabiting fast-flowing streams and rivers with clear, oxygen-rich conditions. These habitats tend to have gravel or sandy substrates for spawning purposes. The species is less likely to be seen in non-flowing, polluted waters, as these conditions do not support its survival and reproduction. Aquatic vegetation is also essential for life, as it provides habitat and feeding grounds. The younger waters of many fish live in aquatic plants.

This species is categorized as Least Concern (LC) according to version 3.1. in IUCN Red List.

3.3.4 Zingel balcanicus *(Karaman, 1937)*

Common Name: Balkan Zingel, Vretenar (Macedonian name)

$$D_1\ \mathrm{VIII}; D_2\ \mathrm{II}\ 12-13; \mathrm{A}\ \mathrm{II}\ 10-11; \mathrm{P}\ 14-15 \tag{25}$$

This species of freshwater fish, belonging to the Percidae family, is native to the Vardar catchment area and is recognized for its limited distribution and conservation concerns. *Zingel balcanicus* generally reaches lengths of 15–20 cm, although some individuals may grow larger. It has a slender, elongated form with a slightly flattened head, typically displaying a grayish-brown hue with darker spots that provide effective camouflage against the riverbed.

Balkan Zingel primarily inhabits fast-flowing, clear rivers characterized by rocky or gravelly bottoms, and mainly found within the Vardar River basin in North Macedonia.

This carnivorous fish primarily feeds on small invertebrates, insects, and various benthic organisms, occasionally preying on smaller fish or larvae. It is well-adapted to swift currents, often staying close to the riverbed to ambush its prey and exhibits increased activity during twilight and nighttime.

Spawning takes place in the spring, with females depositing eggs in shallow, rapidly flowing areas, typically in crevices between rocks or on gravel beds.

According to version 3.1. in the IUCN Red List *Zingel balcanicus* is categorized as Vulnerable (VU).

The main threats for this species are habitat destruction, pollution, dam construction, and competition with invasive species. Conservation initiatives focus on preserving habitats, controlling pollution, and monitoring efforts to maintain healthy populations of this species.

As a predator, *Zingel balcanicus* plays an essential role in regulating smaller invertebrate and fish populations, thereby contributing to the ecological balance of its environment. Ongoing research seeks to deepen the understanding of its ecology and the environmental challenges it encounters, emphasizing population dynamics and effective conservation methods.

3.3.5 Chondrostoma vardarense *(Karaman, 1928)*

Common Name: Vardar Nase, vardarski bojnik, vardarski klen (Macedonian name)

$$\begin{gathered} \text{D III } 8-9;\ \text{A III } 8-10;\ \text{V II } 8;\ \text{P I } 16-17; \\ \text{l.l.}10-60-64-6 \end{gathered} \tag{26}$$

This freshwater fish, belonging to the Cyprinidae family, is native to the Vardar River basin, underscoring its importance in the Balkan region. *Chondrostoma vardarense* typically grows to lengths ranging from 15 to 20 cm, with some specimens reaching up to 25 cm. It features an elongated, cylindrical body characterized by a silvery hue, darker on the dorsal side and lighter on the ventral side. Its mouth is uniquely adapted for scraping algae and other materials from submerged surfaces.

This species is present in clear, swiftly flowing rivers and streams that have gravelly or rocky bottoms.

The primary diet of Vardar Nase consists of algae, periphyton, and other plant matter, though it may also ingest small invertebrates.

This fish is recognized for its tendency to often gather in groups to reduce the risk of predation. Spawning occurs during the spring months, involving migrations to shallow, fast-flowing regions for egg-laying.

According to version 3.1. in the IUCN Red List *Chondrostoma vardarense* is categorized as Near Threatened (NT).

This fish species is threatened by habitat loss, pollution, dam construction, and the introduction of nonnative species. Conservation initiatives focus on safeguarding habitats and preserving the natural flow of rivers.

Chondrostoma vardarense plays a vital role in regulating algae growth and is essential to the riverine ecosystem. Ongoing research examines its ecology, behavior, and the environmental challenges it encounters, emphasizing the necessity of conservation measures for its continued existence.

3.3.6 Salmo macedonicus *(Karaman, 1924)*

Common Name: Macedonian Trout, Makedonska pastrmka (Macedonian name)

$$\begin{gathered} \text{D IV 9, A IV 7} - 8, \text{V II 8, P I 11} - 13, \text{C 19} \\ \text{l.l.21} - 27 - 115 - 122 - 21 - 23 \text{ Ap.Pyl.49} - 74 \end{gathered} \tag{27}$$

This trout species, which is of medium size, is a member of the Salmonidae family and is indigenous to the Balkan Peninsula, where its limited range makes it particularly noteworthy. Adult individuals of Macedonian Trout generally reach lengths between 20 and 40 cm, although some can grow up to 50 cm. The fish exhibits a streamlined body with a coloration that ranges from brownish to olive, featuring distinctive black spots along its sides and a lighter underside.

Salmo macedonicus inhabits cold, well-oxygenated freshwater ecosystems, predominantly within the Vardar River basin, including its tributaries and the mountain lakes in the region.

This species is carnivorous, primarily consuming insects, crustaceans, Macedonian Trout tends to be solitary and territorial, particularly during the spawning season. Spawning generally occurs during the autumn and early winter month, when the fish migrate to appropriate gravel beds for laying eggs.

According to version 3.1. in the IUCN Red List *Chondrostoma vardarense* is categorized as Vulnerable (VU).

Salmo macedonicus is threatened by habitat loss, pollution, overfishing, and climate change. Conservation initiatives are aimed at protecting habitats and restoring natural river ecosystems.

Acting as a top predator, this species plays a crucial role in maintaining the balance of aquatic life and serves as an indicator of water quality and overall ecosystem health. Ongoing research is focused on understanding its ecology and habitat needs, highlighting the urgent need for conservation efforts to ensure its continued survival.

3.3.7 Salmo pelagonicus *(Karaman, 1938)*

Common Name: Pelagonian Trout, Pelagoniska pastrmka (Macedonian name).

This medium-sized trout species is endemic to the Balkan Peninsula, particularly the Crna River basin in North Macedonia. Adults typically reach lengths of 20–30 cm, with some growing larger. The Pelagonian trout features a body with brownish or olive coloration, a lighter belly, and dark spots, particularly pronounced in males during spawning.

Salmo pelagonicus prefers cold, well-oxygenated freshwater environments, it thrives in fast-flowing mountain streams and rivers with rocky or gravelly substrates.

This carnivorous fish feeds on aquatic invertebrates, insects, and small fish, with dietary variations occurring seasonally.

Pelagonian trout are territorial and solitary, particularly during breeding. Spawning occurs in autumn and early winter in shallow, gravelly areas, where females create nests for their eggs.

According to version 3.1. in the IUCN Red List *Chondrostoma vardarense* is categorized as Vulnerable (VU).

This species faces threats from habitat destruction, pollution, overfishing, and climate change. Conservation efforts aim to protect its habitat and manage environmental threats.

As a predator, *Salmo pelagonicus* helps regulate smaller fish and invertebrate populations, maintaining the ecological balance of its environment. Ongoing research is vital for understanding its ecology and developing effective conservation strategies to ensure its survival.

4. Threats to the recovery of threatened fish species

Endemic fish species are at a greater risk due to the vast range of threats influencing their survival. Since they are restricted to small geographical regions, these species show a remarkable sensitivity toward changes in their environment. Their lives are endangered by the following threats:

- Habitat loss and degradation:

  The deforestation leads to the loss of vital habitats, particularly in upper course rivers where there are growing numbers of critically endangered species of trout (*Salmo peristericus*, *Salmo macedonicus*, and *Salmo pelagonicus*).

- Urbanization: Building and other land-use changes around water bodies, especially natural lakes can replace or modify shorelines rendering the environment inhospitable to native fish. This applies to almost all fish species in Macedonian rivers and lakes.

- Agriculture: Runoff entering Prespa Lake and its tributaries acts as a source of contaminants already present in the ecosystem; these are agricultural pesticides, fertilizers, and sediments that worsen water quality causing habitat degradation which influences bio-accumulation rate within fish tissue.

- Dams and water diversion: the construction of high mountain river dams around Prespa and Ohrid Lake is reducing populations of underwater fish which live from these waters. The coastal habitat of Ohrid Lake has been seriously impacted by the diversion into the lake with large amounts of sediment and sludge which came from projects diverting the Sateska River—there are no longer any natural spawning grounds for endemic fish in that part of the lake.

- Pollution: Industrial discharge, Agricultural waste, and Household Waste add hazardous chemicals into rivers and lakes poisoning fish which in turn destroy their habitats. Eutrophication with too much nutrients into the water creates algal blooms in Prespa and Dojran Lakes which use up all oxygen. Then we see hypoxia zones forms where fish cannot live.

- Climate Change: A prominent impact of global warming is often the water temperature change that effects and makes unsuitable habitats for endemic fish species sensitive to changes in this factor.

- Altered precipitation: Changes in the timing, form, and amount of rainfall affect water availability (quantity) and impacts on habitat conditions for native freshwater fish. Overfishing and Exploitation Commercial Fishing: Endemic fishes suffer from overfishing that targets them for food, trade or sport. The Ohrid trout population has dwindled and today it would take many decades to restore their numbers, age structure, and sex ratios back to the historical conditions. IUU Fishing—Macedonia Southern Waters of the Republic, Greece Unsustainable fishing practices in Macedonia waters severely reduce wide endemic fish species populations. This is a common practice because, for the most part, there may be no one to check or regulate that sort of thing.

- Invasive Species Predation: A nonnative species in a habitat that preys on endemic fish can reduce their population. It is especially noticeable at Prespa Lake with the intentional introduction of *Lepomis gibbosus*, which outcompetes for food and habitat resulting in a drastic decline in abundance, nearly all endemic fish species present within its drainage. Carassius impact on native populations in Lake (Dojran). In the case of Dojran Lake, it is known that long-term stable community structure.

- Tourism and recreation: Increased human activity can cause negative changes to fish habitats and spawning grounds.

- Reduction of the water level in lakes and rivers: Prespa Lake and Dojran Lake are facing a reduced water level that significantly affects the survival of endemic fish species. With the reduction of the coastal zones in the lakes, the spawning grounds of the fish are destroyed, affecting all the populations of these fish, and disrupting their growth and development.

5. Conclusions

1. Macedonian Fresh Waters belong to two ecoregions, namely: Ecoregion 420 (watershed of the southeastern Adriatic, Lake Ohrid, and Lake Prespa) and Ecoregion 422 (watershed of the Vardar River with Lake Dojran and the Struica River)

2. In Ecoregion 420 (watershed of the southeastern Adriatic, in Lake Ohrid) 14 endemic species have been described, namely are: *Alburnoides ohridanus*, *Alburnus scoranza*, *Barbatula sturanyi*, *Barbus rebeli*, *Cobitis ohridana*, *Gobio ohridanus*, *Pachychilon pictum*, *Pelasgus minutus*, *Rutilus ohridanus*, *Salmo aphelios*, *Salmo balcanicus*, *Salmo letnica*, *Salmo lumi* and *Salmo ohridanus.*

3. In Ecoregion 420 (watershed of the southeastern Adriatic, in Lake Prespa) 10 endemic species have been described, namely are: *Alburnoides prespensis*, *Alburnus pelvic*, *Barbus prespensis*, *Chondrostoma prespense*, *Cobitis meridionalis*, Economidichthis pigmaeus, *Pelasgus prespensis*, *Rutilus prespensis*, *Salmo peristericus*, and *Squalius prespensis*.

4. In Ecoregion 422 (watershed of the Vardar River with Lake Dojran and the Strumica River) seven endemic species have been described: *Alburnus*

macedonicus, *Pachychilon macedonicum*, *Zingel balcanicus*, *Chondrostoma vardarense*, *Gobio bulgaricus*, *Salmo macedonicus*, and *Salmo pelagonicus*.

5. Thirty-one endemic species of fish have been found in the Macedonian freshwater waters, at least for the time being, which are represented in the waters of the Republic of Macedonia or in the countries belonging to the river basin which also belongs to the Republic of North Macedonia.

6. In the research that will be carried out in the next period, new species will certainly be discovered, which I will add to the list of new and even endemic species e.g., *Economidichthys pygmaeus*, which was recorded in Lake Prespa but was not established in list of endemic species for Lake Prespa.

Author details

Trajche Talevski[1*], Marina Talevska[2] and Aleksandra Talevska Leshoska[3]

1 Independent Researcher, Ohrid, R.N. Macedonia

2 Hydrobiological Institute, Ohrid, R.N. Macedonia

3 PSI Biomed Lab, Ohrid, R.N. Macedonia

*Address all correspondence to: tratal2001@gmail.com

References

[1] Talevski T. Peculiarities of the population of moranec *Pachychilon pictum* (Heckel & Kner, 1858) from Lake Ohrid [doctoral thesis]. Skopje Macedonia. Macedonian language: Faculty of Natural Sciences and Mathematics; 2003

[2] Bănărescu PM. Einige Fragen zur Herkunft und Verbreitung der Süsswasserfischfauna der europäisch-mediterranen Unterregion. Archiv fuer Hydrobiologie. 1960;**57**(1/2):16-134

[3] Станковиќ С. Охридското Езеро и неговиот жив свет. Скопје: Култура; 1959. 421 pages Macedonian Macedonian language

[4] Stankovic S. The Balkan Lake Ohrid and its Living World. Monographiae Biologiae. Vol. IX, 357 pp. Den Haag: Uitgeverij Dr. W. Junk; 1960

[5] Vuković T, Ivanović B. Slatkovodne ribe Jugoslavije. Zemaljski muzej Bosne i Hercegovine u Sarajevu, Prirodnjačko Odjeljenje. Posebno izdanje. 1971. 268 p. [In Serbian]

[6] Karaman S. Pisces Macedoniae. Split. 1924

[7] Kottelat M, Freyhof J. Handbook of European Freshwater Fishes. Berlin, Germany. ISBN: 978-2-8399-0298-4. 646 p: Kottelat, Cornol, Switzerland, Freyhof; 2007

[8] Stefanovic D. Rasna i ekoloska ispitivanjana Ohridskim salmonidama [Doctoral thesis (published posthumously)]. Serbian language 1948

[9] Karaman S et al. Bulletin de la Société Scientific de Skoplje. 1928;**6**:147, 170-176. In Russian, German summary, 176

[10] Abell R, Thieme L, Revenga C, Bryer M, Kottelat M, Bogutskaya N, et al. Freshwater ecoregions of the world: A new map of biogeographic units for freshwater biodiversity conservation. Bioscience. 2008;**58**(5):403-414. DOI: 10.1641/B580507

[11] The IUCN Red List Categories and Criteria. Available from: https://www.iucnredlist.org/

[12] Milošević D, Talevski T. Conservation status of native species in Natural Lakes of Drim system (Prespa, Ohrid and Skadar Lake) and dangers of commercial fishing. Bulgarian Journal of Agricultural Science. 2015;**21** (Supplement 1):61-67

[13] Talevski T, Milosevic D, Talevska A. Anthropogenic influence and conservation status of autochthonous fish fauna from Lake Ohrid. In Conference of Water Observation and Information System for Decision Support, Balwois 2010. 2010. Available from: http://balwois.com/balwois/administration/full_paper/ffp-1587.pdf

[14] Talevski T, Milosevic D, Talevska A. Anthropogenic influence and conservation status of autochthonous fish fauna from Lake Prespa. In: Conference of Water Observation and Information System for Decision Support, Balwois 2010. 2010. Available from: http://balwois.com/balwois/administration/full_paper/ffp-1588.pdf

[15] Talevski T, Milosevic D. Biodiversity and conservation status of the fish in the great natural lakes (Dojran Lake, Prespa Lake and Ohrid Lake) of the Macedonia. Agricultural Academy. Bulgarian Journal of Agricultural Science. 2015;**21** (Supplement 1):61-67

[16] Talevski T, Talevska M. Biodiversity and conservation status of the fish and

qualitative composition of the macrophytic vegetation in the Lake Dojran, Republic of Macedonia. In: 8th International Conference "Water & Fish"—Conference Proceedings. 2018. pp. 448-452

[17] Trajchevski B, Spirkovski Z, Ilikj-Boeva, Talevski T. An alien species or another perspective to the freshwater gobies puzzle: A new finding in Lake Prespa. Turkish Journal of Zoology. 2020;**44**(44):542-547. DOI: 10.3906/zoo-2008-20